CHILDREN'S GUIDE TO THE NIGHT SKY

Tom Kerss

Published by Collins
An imprint of HarperCollinsPublishers
Westerhill Road
Bishopbriggs
Glasgow
G64 2QT

HarperCollinsPublishers
Macken House, 39/40 Mayor Street Upper, Dublin 1, D01 C9W8, Ireland

collins.co.uk

First published 2024

© HarperCollinsPublishers 2024

Collins® is a registered trademark of HarperCollinsPublishers Ltd.

In association with the National Space Centre

Text © Tom Kerss
Cover illustrations by Steve Evans

Publisher: Michelle I'Anson
Project leader: Rachel Allegro
Design: Kevin Robbins and James Hunter
Cover: Steve Evans, Kevin Robbins and James Hunter
Illustration: Steve Evans
Production: Ilaria Rovera

All rights reserved. No part of this publication may be reproduced, stored in a retrieval system, or transmitted, in any form or by any means, electronic, mechanical, photocopying, recording or otherwise without the prior permission in writing of the publisher and copyright owners.

The contents of this publication are believed correct at the time of printing. Every care has been taken in the preparation of this book. However, the publisher can accept no responsibility for errors or omissions, changes in detail given or for any expense or loss thereby caused.

The publisher does not warrant that any website mentioned in this title will be provided uninterrupted, that any website will be error free, that defects will be corrected, or that the website or the server that makes it available are free of viruses or bugs. For full terms and conditions please refer to the site terms provided on the website.

A catalogue record for this book is available from the British Library.

ISBN 9780008700331

Printed by Martins The Printers, UK

10 9 8 7 6 5 4 3 2

This book contains FSC™ certified paper and other controlled sources to ensure responsible forest management.

For more information visit: www.harpercollins.co.uk/green

Collins

CHILDREN'S GUIDE TO THE NIGHT SKY

100 Things to See in Space

Tom Kerss

CONTENTS

100 THINGS TO SEE	**8**
STARGAZING	**14**
STAR CHARTS	**46**
THE SOLAR SYSTEM	**84**
DEEP SKY	**120**

UNUSUAL SIGHTS	**140**
ASTROPHOTOGRAPHY	**148**
RESOURCES	**162**
GLOSSARY	**164**
INDEX	**170**

INTRODUCTION

You are about to embark on an incredible journey of discovery into the night sky. It is full of remarkable things waiting to be uncovered.

Using just your eyes, a pair of binoculars or a telescope, you can explore the Universe from your own home. You'll be following in the footsteps of the greatest astronomers in history!

This book contains everything you need to know to start your own astronomy adventure. Astronomy is the study of stars, planets, comets, galaxies, space and anything else outside of the Earth and its atmosphere. Whether you want to point out the star constellations after sunset, see storms on other planets or gaze across the cosmos at distant galaxies, you'll find the tools and tips you need in the pages ahead.

I hope this book leaves you ready for a lifetime of exploration, whether you choose to keep on stargazing as a hobby or go even further into an exciting astronomy career of your own.

I wish you very clear skies!

Tom Kerss

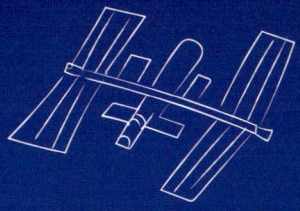

CHALLENGE YOURSELF

There are 100 things to see listed on pages 10–13. You can find some of these tonight! However, you will have to wait for others to appear. Challenge yourself to find as many of them as you can!

ASK FOR HELP WHEN YOU NEED IT

Stargazing is really fun but it takes patience and practice! Even scientists and astronomers are learning new things all the time! Start by getting familiar with the stars and constellations and have a look for objects in our Solar System. Then, when you're ready, you can move onto the more challenging Deep Sky! Ask a grown-up if you need help and take a look at the online resources on page 162.

BE SAFE!

When going stargazing, always take a grown-up with you, wrap up warm and remember to take snacks and drinks. If you're observing the Sun, never look directly at it or point binoculars or a telescope at it.

WHAT IS THE NATIONAL SPACE CENTRE?

The National Space Centre is an award-winning visitor attraction and educational charity. Home to the UK's largest planetarium, you can journey through the wonders of the Universe and learn about space exploration. We'll be helping to answer questions all about astronomy through the book. Find out more at **spacecentre.co.uk**

100 THINGS TO SEE

Challenge yourself to see how many of these things you can spot in the night sky.

ASTERISMS   PAGE

- ✓ Plough 24, 28, 29, 130
- ○ Northern Cross 25
- ○ Teapot 25, 62, 63, 124, 129
- ○ Summer Triangle 26, 62, 63, 125
- ○ Winter Hexagon 26, 78, 79
- ○ Coathanger 27, 73
- ○ Keystone 27, 64, 65, 128
- ○ Sickle 54, 55
- ○ Great Square of Pegasus. 68–71
- ○ Circlet. 70, 71
- ○ Kemble's Cascade. 66, 67
- ○ Orion's Belt. 79
- ○ Kids 80, 81

STARS

- ○ Polaris 28, 29, 50, 58, 59, 66, 74, 75
- ○ Sirius. 26, 35, 78, 79
- ○ Arcturus. 35, 52–55
- ○ Capella 26, 35, 78–81
- ○ Betelgeuse. 35, 78, 79
- ○ Merak. 28, 51
- ○ Dubhe. 28, 51
- ○ Deneb. 25, 26, 60–63, 72, 73, 125
- ○ Vega. 26, 35, 62, 63, 131
- ○ Altair 26, 62, 63
- ○ Albireo 25, 60–63, 130
- ○ Antares 62, 63, 129

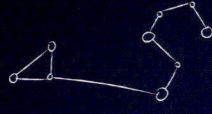

- Almach 68, 69, 131
- Algol 68, 69
- Algieba 76, 77
- Mizar 28, 51, 130
- Castor 57, 78, 79, 131
- Pollux 26, 57, 78, 79

CONSTELLATIONS

- Virgo 18, 30, 31, 54, 55, 64
- Ursa Major . . . .
  . . . .19, 28, 50–54, 56–59, 64, 66, 67, 74–78
- Ursa Minor . . . .
  . . . .19, 28, 50, 52, 56, 58, 59, 64, 66, 74, 75
- Orion 19, 26, 78, 79, 116, 133
- Perseus . . . 19, 58, 59, 66–69, 74, 78, 80, 127
- Gemini . . . 26, 30, 31, 56, 57, 76, 78, 79, 131
- Taurus 26, 30, 31, 68, 78–81, 126
- Cancer 30, 31, 56, 57, 76, 77, 78, 127
- Leo 30, 31, 54–56, 64, 76, 77, 138, 142
- Scorpius 30, 31, 62, 63, 129
- Aquarius 30, 31, 70, 71
- Hydra 54, 55, 76
- Boötes 50, 52–54, 58, 62, 64
- Cygnus . . . .
  . . 25, 26, 60–63, 66, 70, 72, 73, 125, 130, 133
- Lyra 26, 58, 60, 62, 63, 72, 131
- Aquila 26, 60, 62, 63, 72
- Hercules . 27, 52, 58–60, 62, 64, 65, 72, 128
- Draco . 50, 52, 58–60, 62, 64–66, 72, 74, 75
- Cassiopeia . . . .
  . . . 50, 58, 59, 66–70, 72, 74, 75, 78, 80, 127

- Perseus . . . 19, 58, 59, 66–69, 74, 78, 80, 127
- Pegasus. 60, 68, 70–73, 80, 129
- Vulpecula. 27, 60, 72, 73
- Sagitta 27, 60, 72, 73
- Cepheus 50, 58, 60, 66, 68, 70, 72–74
- Andromeda . . 60, 66, 68–70, 72, 73, 80, 131
- Auriga. 26, 56, 58, 68, 74, 76, 78–81
- Taurus. 26, 30, 31, 68, 78–81, 126

SOLAR SYSTEM

10 things to find on the Moon
- Copernicus crater 90, 91
- Sinus Iridum 90, 91
- Tycho crater 90, 91
- Clavius crater. 90, 91
- Rupes Recta 90, 91
- Aristarchus crater 90, 91
- Plato crater. 90, 91
- Apennines Mountains. 90, 91
- Palus Somni 90, 91
- Theophilus, Cyrillus
  and Catharina craters 90, 91
- Eclipse (solar and lunar) 94–99
- Mercury. 100, 101
- Venus 102, 103
- Mars. 63, 104, 105
- Jupiter 106, 107
- Saturn. 108, 109
- Uranus 110, 111
- Neptune 110, 111

- Asteroid 112, 113
- Comet 114–116
- Meteor shower 116, 117
- International Space Station (ISS) 118

DEEP SKY OBJECTS

- Milky Way . . 38, 49, 58–63, 78, 79, 122–125
- Coma Star Cluster 54, 55
- M104 – Sombrero Galaxy . . 54, 55, 137–139
- M44 – Beehive Cluster . . . 56, 57, 76, 77, 127
- M51 – Whirlpool Galaxy 50–53, 136
- M92 (Globular Cluster) 64, 65
- C6 – Cat's Eye Nebula 58, 59
- M27 – Dumbbell Nebula . 60, 61, 72, 73, 135
- M71 – Angelfish Cluster 73
- C14 – Double Cluster 68, 69, 127
- IC 1805 – Heart Nebula 68, 69
- M31 – Andromeda Galaxy 68, 69, 136
- M42 – Orion Nebula 78, 79, 133
- M45 – Pleiades Star Cluster . . . 78, 79, 126
- C41 – Hyades Star Cluster 80, 81, 126
- Cygnus Star Clouds 73, 125
- M11 – Wild Duck Cluster 62, 63, 127
- M13 – Hercules Cluster 64, 65, 128
- Leo Triplet 54, 55, 138, 139
- Virgo Galaxy Cluster 138, 139, 143

STARGAZING

STARGAZING THROUGHOUT HISTORY

What do you think ancient people thought about the night sky?

Our ancestors have admired the night sky for at least tens of thousands of years. There is evidence of ancient star charts carved into rocks dating back over 30,000 years.

We can only guess what ancient people thought about the night sky because there are no written records from this time to tell us. Were these people afraid if unexpected things happened, like if the Sun disappeared behind the Moon? Were they scared if they saw a comet burning through the night sky? Were they fascinated, or even comforted, by the stars?

THE SEASONAL SKY

Stargazing has always had an important role in people's lives. Several thousand years ago, farmers began to use the stars as a calendar. The stars helped guide them when to sow and harvest crops so that they could plan and organise their lives. This meant they could stay in one place and grow food for their community on farms, rather than travelling around and hunting for food to eat. Once people could live permanently in one place, the first civilisations began to develop.

We wouldn't be here today if our ancestors hadn't used what they knew about the night sky to make farming possible.

Our knowledge of astronomy has developed a lot in the last few hundred years, with the invention of the telescope and all the science that followed. The stars themselves are no less fascinating, though, and the stories our ancestors told about them thousands of years ago still live on today.

STORIES
IN THE NIGHT SKY

When people looked at the stars thousands of years ago, they saw that they could be linked together in patterns or shapes. These groups of stars are known as constellations. Some of the earliest constellations in use today, such as Taurus and Leo, are probably over 15,000 years old, perhaps even older.

Over the years, people around the world have told stories about the star patterns. Some of these stories have survived. For example, we still know many of the stories that people in Ancient Greece told about the stars. Almost every constellation that can be seen from Europe has a Greek story. These stories tell us information about the heroes, animals and objects that were important to people who lived long ago.

VIRGO

The stars of the constellation Virgo used to rise in the east before sunrise in late summer, just ahead of the time to harvest the grain. Because of this, Virgo is said to be the goddess of farming and agriculture and is shown holding an ear of wheat.

URSA MAJOR AND URSA MINOR

Ursa Major (Great Bear) and Ursa Minor (Little Bear) were originally said to be humans called Callisto and Arcas. The Greek god, Zeus, turned them both into bears to protect them. He flung them into the sky, stretching out their tails.

URSA MINOR

URSA MAJOR

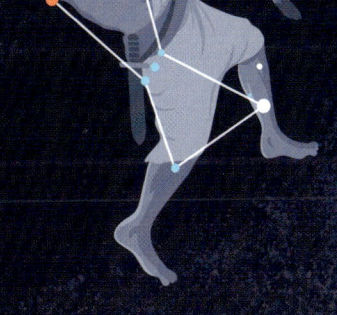

ORION

According to Greek mythology, Zeus also sent a scorpion to kill the mighty hunter, Orion, for being boastful. Zeus then placed both Orion and the scorpion in the sky, to remind us not to be boastful.

PERSEUS

Perseus features in the stories of many constellations across the sky. One story tells of how he rescued the princess Andromeda from a ferocious sea monster. He arrived on Pegasus, his flying horse, just in time to save the day.

19

CONSTELLATIONS AND THEIR MEANINGS

Over time, more and more constellations have been identified, and now every star in the sky belongs to one. In 1930, astronomers from all over the world agreed on a list of 88 'modern' constellations.

Name (Latin)	Abbreviation	Meaning
Andromeda	And	Princess of Ethiopia
Antlia	Ant	Air pump
Apus	Aps	Bird of Paradise
Aquarius	Aqr	Water/cup bearer
Aquila	Aql	Eagle
Ara	Ara	Altar
Aries	Ari	Ram
Auriga	Aur	Charioteer/goat shepherd
Boötes	Boo	Herdsman
Caelum	Cae	Chisel
Camelopardalis	Cam	Giraffe
Cancer	Cnc	Crab
Canes Venatici	CVn	Hunting dogs
Canis Major	CMa	Big dog
Canis Minor	CMi	Little dog
Capricornus	Cap	Sea goat
Carina	Car	Keel of the Argonauts' ship
Cassiopeia	Cas	Queen of Ethiopia
Centaurus	Cen	Centaur
Cepheus	Cep	King of Ethiopia
Cetus	Cet	Sea monster (whale)
Chamaeleon	Cha	Chameleon

? DID YOU KNOW?

On star charts, constellations are sometimes abbreviated (shortened). For example, 'Circinus' is abbreviated to 'Cir'.

Name (Latin)	Abbreviation	Meaning
Circinus	Cir	Compasses
Columba	Col	Dove
Coma Berenices	Com	Queen Berenice's hair
Corona Australis	CrA	Southern crown
Corona Borealis	CrB	Northern crown
Corvus	Crv	Crow
Crater	Crt	Cup
Crux	Cru	Cross (southern)
Cygnus	Cyg	Swan
Delphinus	Del	Dolphin
Dorado	Dor	Swordfish
Draco	Dra	Dragon
Equuleus	Equ	Little horse
Eridanus	Eri	River
Fornax	For	Furnace
Gemini	Gem	Twins
Grus	Gru	Crane
Hercules	Her	Hercules, son of Zeus
Horologium	Hor	Clock
Hydra	Hya	Sea serpent
Hydrus	Hyi	Water snake
Indus	Ind	Indigenous person

CONSTELLATIONS AND THEIR MEANINGS

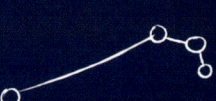
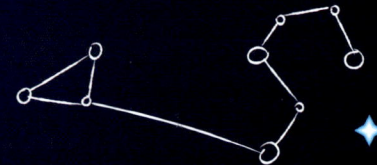

Name (Latin)	Abbreviation	Meaning
Lacerta	Lac	Lizard
Leo	Leo	Lion
Leo Minor	LMi	Little lion
Lepus	Lep	Hare
Libra	Lib	Balance
Lupus	Lup	Wolf
Lynx	Lyn	Lynx
Lyra	Lyr	Lyre or harp
Mensa	Men	Table mountain
Microscopium	Mic	Microscope
Monoceros	Mon	Unicorn
Musca	Mus	Fly
Norma	Nor	Carpenter's level (type of tool)
Octans	Oct	Octant (tool for navigation)
Ophiuchus	Oph	Holder of serpent
Orion	Ori	Orion, the hunter
Pavo	Pav	Peacock
Pegasus	Peg	Pegasus, the winged horse
Perseus	Per	Perseus, the hero
Phoenix	Phe	Phoenix
Pictor	Pic	Easel
Pisces	Psc	Fishes

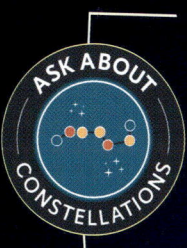

DO WE SEE THE SAME STARS FROM EVERYWHERE ON EARTH?

The stars you see depend upon where you are in the world and the time of year. Some of them can only be seen in the skies of the northern hemisphere, while others are seen in the southern hemisphere.

Name (Latin)	Abbreviation	Meaning
Piscis Austrinus	PsA	Southern fish
Puppis	Pup	Stern of the Argonauts' ship
Pyxis	Pyx	Compass of the Argonauts' ship
Reticulum	Ret	Net
Sagitta	Sge	Arrow
Sagittarius	Sgr	Archer
Scorpius	Sco	Scorpion
Sculptor	Scl	Sculptor's tools
Scutum	Sct	Shield
Serpens	Ser	Serpent
Sextans	Sex	Sextant (tool for navigation)
Taurus	Tau	Bull
Telescopium	Tel	Telescope
Triangulum	Tri	Triangle
Triangulum Australe	TrA	Southern triangle
Tucana	Tuc	Toucan
Ursa Major	UMa	Great bear
Ursa Minor	UMi	Little bear
Vela	Vel	Sails of the Argonauts' ship
Virgo	Vir	Virgin
Volans	Vol	Flying fish
Vulpecula	Vul	Little fox

ASTERISMS

As well as the 88 constellations there are some other familiar star patterns in the night sky, known as asterisms.

Asterisms are very popular with stargazers because they are made up of similarly bright stars which are easy to pick out. Some asterisms are larger than a single constellation.

THE PLOUGH

The Plough (sometimes known as the Big Dipper) is formed from the brightest stars of Ursa Major (the Great Bear). It is one of the most famous asterisms in the night sky.

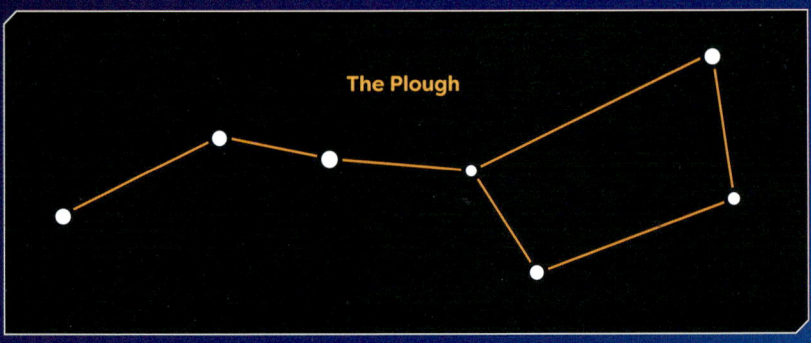

The Plough

? DID YOU KNOW?

The Plough is used to help astronomers and stargazers find true north. See page 28 for how to find true north.

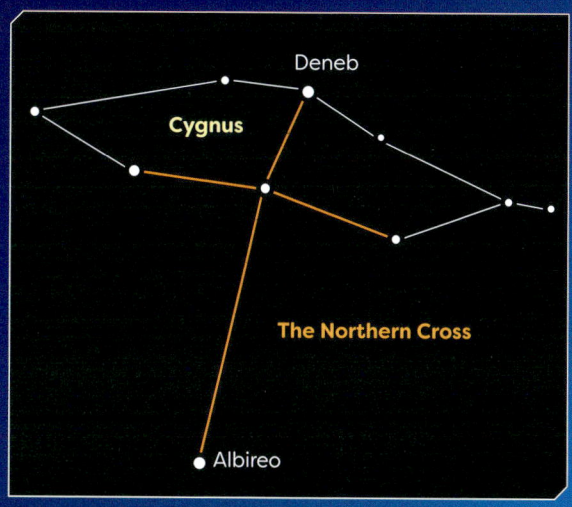

THE NORTHERN CROSS

The five brightest stars in the constellation Cygnus (the swan) form a large cross, or kite. The star at the top of the cross is Deneb, the tail of the swan. The star at the bottom is Albireo, the swan's beak.

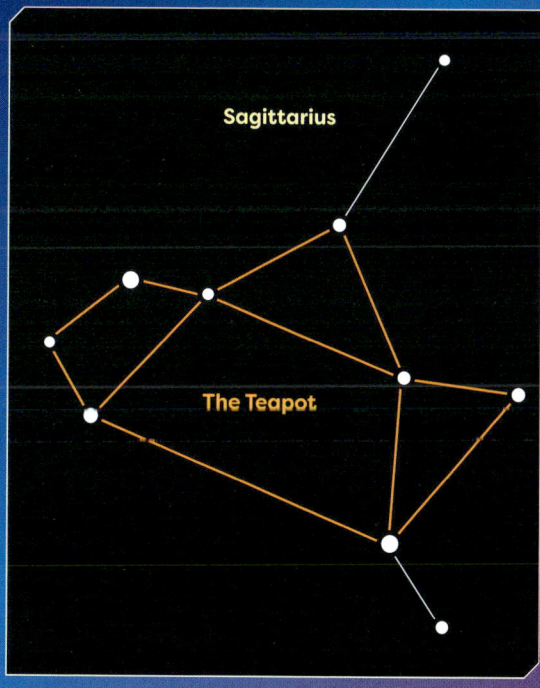

THE TEAPOT

Sagittarius is an archer-centaur, whose bow and arrow looks similar to a teapot. The spout of the teapot is made up of the bow and arrow itself, and the handle is formed from the archer's arm drawing back the arrow. The asterism is visible in summer evening skies in the northern hemisphere.

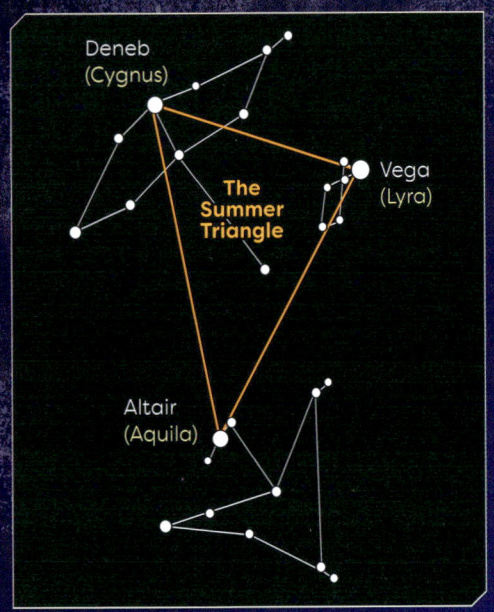

THE SUMMER TRIANGLE

In the summer, we can see Deneb in the constellation Cygnus (the swan), forming part of a very large triangle spread across several constellations. The other bright stars are Vega in Lyra (the harp) and Altair in Aquila (the eagle).

THE WINTER HEXAGON

The enormous Winter Hexagon is even bigger than the Summer Triangle. It is made up of six dazzling winter stars. At the top is Capella in Auriga (the charioteer). From there, going clockwise, you'll find Aldebaran in Taurus, Rigel in Orion, Sirius in Canis Major, Procyon in Canis Minor and Pollux in Gemini.

THE COATHANGER

The constellation Vulpecula (little fox) is home to a small asterism called the Coathanger. See if you can count all ten stars through your binoculars or telescope!

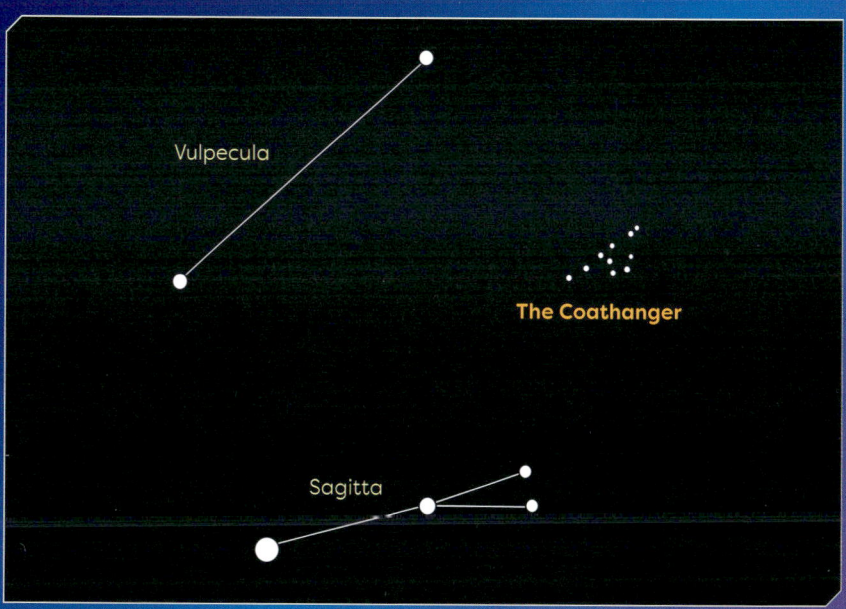

THE KEYSTONE

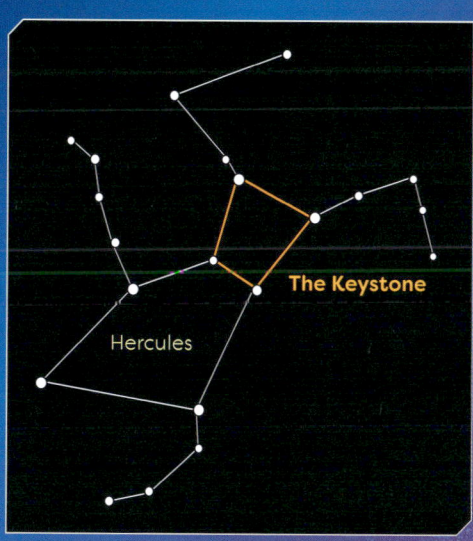

The centre of the constellation Hercules is called the Keystone – four stars squashed into a square shape. It's handy to find this first, before tracing out his sprawling arms and legs in the sky.

FINDING TRUE NORTH

When you go outside at night, the best way to understand where you are is to find true north using the stars. Luckily, this is very easy to do!

In the northern hemisphere, Polaris (known as the North Star) appears to stay fixed in the sky. This is because it's positioned above the North Pole. As the Earth spins on its axis, Polaris doesn't appear to spin around like the other stars, so this star helps you to find true north.

FIND POLARIS

On a clear night, look for a pattern of stars called the Plough (also known as the Big Dipper). It is made up of seven stars (though if you look closely, you will see that Mizar has a little companion star nearby).

Two of the stars, Merak and Dubhe, are known together as 'the pointers'. Trace a line between these stars and follow it into the sky. The next bright star you come to is Polaris.

The stars of the Plough belong to Ursa Major (the great bear), but Polaris is at the tip of the tail of Ursa Minor (the little bear).

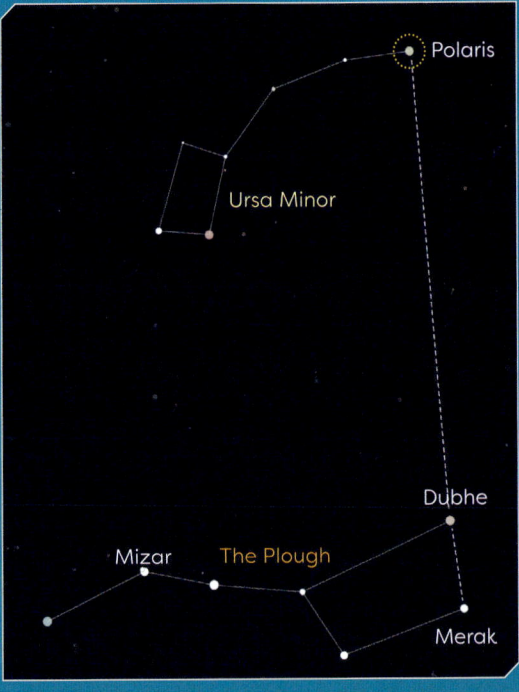

If you could stand at the North Pole, Polaris would be right above your head. That means that when you look towards Polaris from anywhere else in the northern hemisphere, you are looking in the direction of true north. Now that you know how to find true north, you can navigate around the sky using the star charts in this book.

❓ DID YOU KNOW?

The Plough can be seen all year round, but its position is different during each of the four seasons.

CONSTELLATIONS OF THE ZODIAC

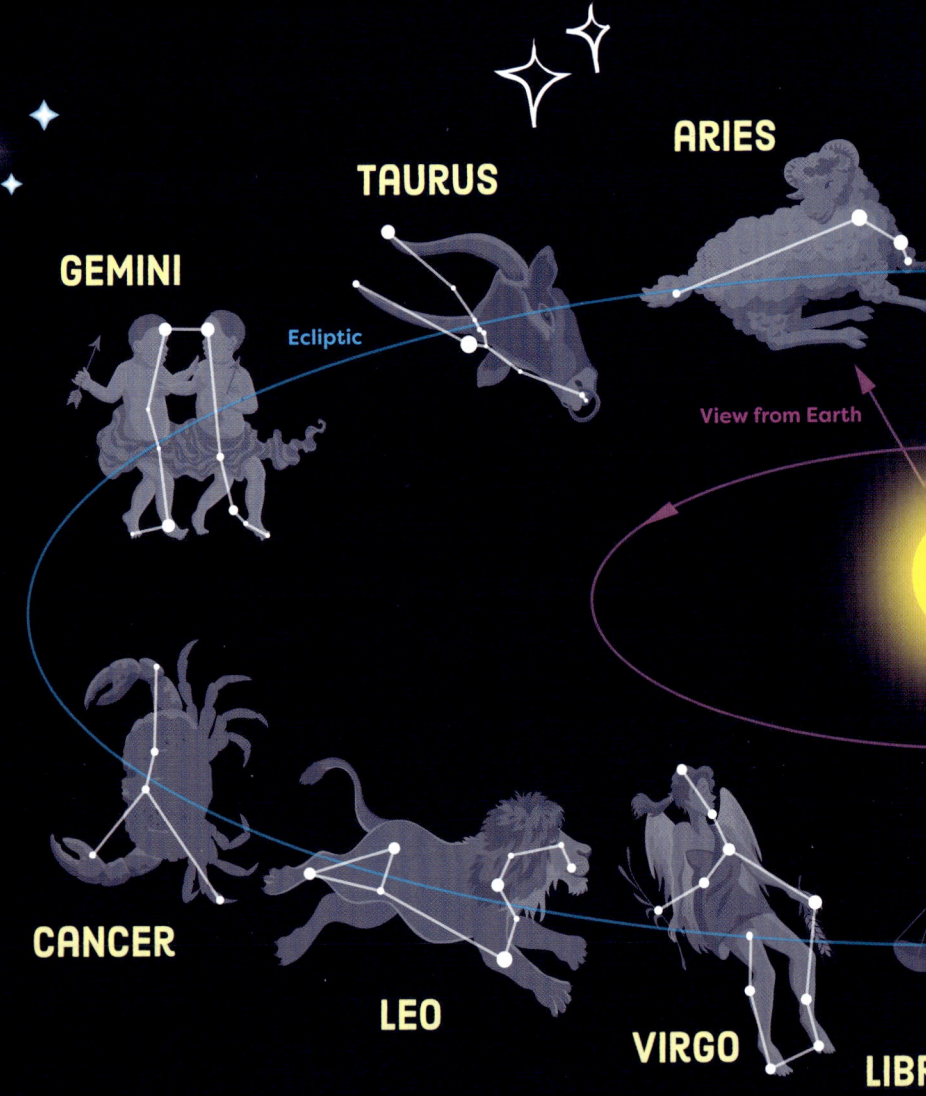

The Sun appears to follow a path through the sky. This imaginary path or line is known as the ecliptic.

As the Sun moves along the ecliptic during the year, it passes through the twelve constellations of the zodiac. These constellations were used by ancient people as a calendar, dividing the year into twelve equal sections.

The sun also passes through a thirteenth constellation called Ophiuchus, but it is not classed as a member of the zodiac.

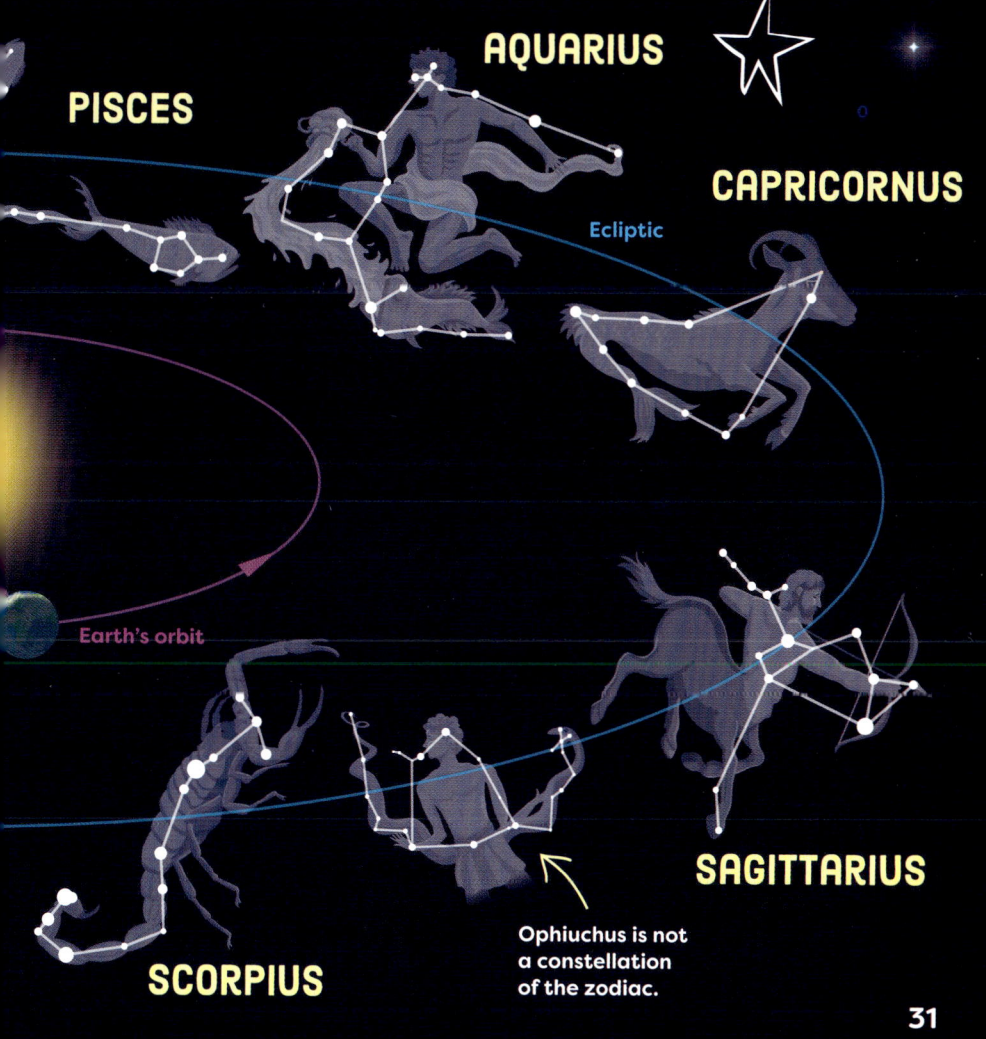

Ophiuchus is not a constellation of the zodiac.

THE EARTH'S SEASONS

The constellations you see at night change with the seasons. This is because the Earth is moving around the Sun, so you are looking at a different part of space.

As the Earth moves around the Sun, it is tilted on its axis. This means that different parts of Earth receive different amounts of sunlight throughout the year. This causes the seasons. The northern hemisphere leans towards the sun in summer and away from it in winter.

When the Earth's north pole has its maximum tilt (23.5°) towards the Sun, it is called the summer solstice. It is the first day of summer and we experience the most daylight.

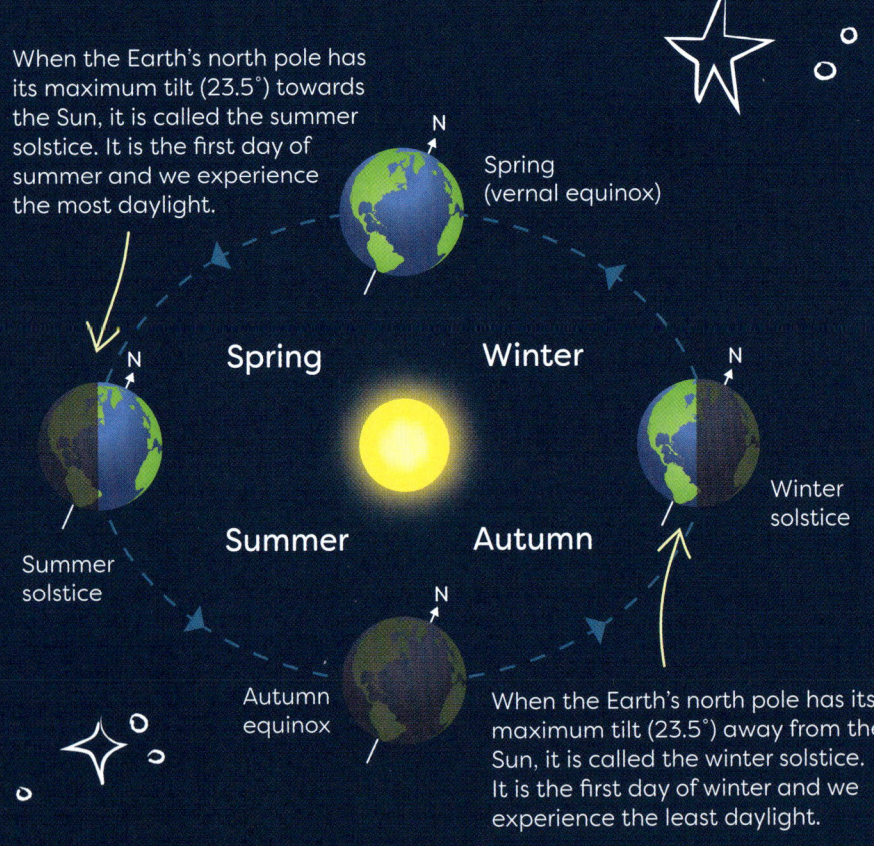

Spring (vernal equinox)

Spring

Winter

Summer

Autumn

Summer solstice

Winter solstice

Autumn equinox

When the Earth's north pole has its maximum tilt (23.5°) away from the Sun, it is called the winter solstice. It is the first day of winter and we experience the least daylight.

An equinox is when the Sun appears directly above the equator. This happens twice a year – on the first day of spring and the first day of autumn. The spring equinox is sometimes called the vernal equinox. On these two days the hours of daylight and darkness are equal.

Winter nights offer more hours of darkness to enjoy spectacular constellations like Orion and Taurus. Summer nights don't get as dark because the Sun is not as far below the horizon. However, during the summer we can see the brightest parts of the Milky Way.

DATES

Spring (vernal) equinox	19–21 March
Summer solstice	20–22 June
Autumn equinox	21–24 September
Winter solstice	20–23 December

STAR BRIGHTNESS AND COLOURS

Looking at the night sky, you can't fail to notice that some stars appear to be brighter than others. One reason for this is that the stars are at different distances from us. The closer they are, the brighter they will appear. But stars really do vary in brightness too. The range is incredible!

The brightest stars ever found are millions of times brighter than the Sun! On star charts, stars that appear bright to our eyes are shown as larger dots, and fainter stars are shown as smaller dots.

VIBRANT COLOURFUL STARS

Stars shine in a variety of colours depending on their different temperatures. You are probably used to thinking of red things as hot and blue things as cool, but with stars it's the other way around. The hottest stars are blue and the coolest stars are red. Red stars are cold compared to other stars, but they are still very hot to us!

Astronomers group stars into different categories according to temperature. From hottest to coolest, they are:

Hottest **O B A F G K M** Coolest

Although star colours appear vibrant in photographs, they are not as easy to see with your eyes. What astronomers describe as a red star, actually looks orange-red to our eyes. Blue stars appear even less colourful to us.

STAR COLOURS AND HOW THEY APPEAR

Category	Labelled colour	How they appear to us
O	Blue	Blue
B	Blue-white	Blue-white
A	White	Pale blue-white
F	Yellow-white	White
G	Yellow	Pale yellow-white
K	Orange	Yellow-orange
M	Red	Orange-red

For example, the Sun is a G-type star.

THE TEN BRIGHTEST STARS IN THE NIGHT SKY

#	Star name	Constellation	Category
1	Sirius	Canis Major	A
2	Canopus	Carina	A
3	Alpha Centauri	Centaurus	G
4	Arcturus	Boötes	K
5	Vega	Lyra	A
6	Capella	Auriga	G
7	Rigel	Orion	B
8	Procyon	Canis Minor	F
9	Betelgeuse	Orion	M
10	Achernar	Eridanus	B

Sirius is the brightest star in the night sky.

CHOOSE A STARGAZING SITE

It's important to know how to choose a great spot for stargazing and what to take with you. Unless you are very lucky, you probably can't see very much of the night sky through your window, even if you turn off the lights!

Imagine the night sky is a big stage – to enjoy it, you'll need to find somewhere with as few things blocking your view as possible.

Usually, the things that stop you being able to see the stars are buildings and trees crowding the horizon, so open spaces make the best stargazing spots. If you have your own garden, you can practise finding the constellations there before heading out into the countryside.

You also need to avoid looking into any bright lights, so turn off any security lights in your garden and be sure you aren't too close to a road where car headlights might dazzle you.

WHAT TO BRING

A good stargazing plan can help you make the most of your 'night out'. Aside from knowing what to look out for, here are some things to bring with you:

Appropriate clothing

It's easy to lose yourself in the magic of the night sky and forget the temperature, but you may be standing outside in the cold for a long time, so make sure you have warm clothes with you.

A red-light torch

Using a torch that creates red light will help you to see in the dark without spoiling your night vision. See page 41 for how to make one.

A notepad and pencil

You can take notes about your observations, record the weather, time and date, and what you saw. You can also make sketches.

Binoculars or telescope (optional)

If you have binoculars or a telescope, don't forget to take them along. Make sure you know how to work them before going out into the dark!

A camera (optional)

If astrophotography interests you, have your camera ready to go, with fully charged batteries!

AVOID LIGHT POLLUTION

Often, the lights we use at night are not well designed, and we waste energy sending light upward into the sky where it is not needed. This is called light pollution.

Around the world, most people live in cities and towns where light pollution is a problem for stargazing. It's not impossible to see the stars in these areas, but the sky is far less impressive than it would appear from the countryside.

DID YOU KNOW?

On a clear, dark night our galaxy, the Milky Way, can be seen as a faint band of light across the sky. But in cities, the Milky Way is impossible to see.

Use the light pollution map in the 'Resources' section to figure out the darkest places near you. Even if you can't go to one of these places, you can still explore the Solar System as the bright planets and Moon are visible in light polluted skies.

MOONLESS SKIES ARE DARKER

To astronomers, the Moon is actually a source of light pollution! It reflects the Sun's light into the sky and becomes a nuisance for stargazers when it is close to being a full moon.

Every 29 to 30 days, there is a period when the Moon is not easily visible (when it's a new moon) and the sky is at its darkest throughout the night. This is the best time to hunt for the faintest objects in the night sky.

Make sure you check the phases of the Moon when you plan to go stargazing. Have a look at page 93 for more on moon phases.

MAKE THE MOST OF YOUR EYES

Your eyes are amazing! They give you sharp, colourful views of the world around you, and focus and adapt to changing distances and brightness.

Unfortunately, though, your eyes aren't perfectly engineered to see in the dark, so you'll have to learn to make the most out of them.

When the light level is low, the pupils of your eyes dilate (get bigger) to let more light in. This is one of the ways your eyes adapt to the dark.

It takes about 30 minutes to adapt to the dark, but just a fraction of a second to lose it if you accidentally gaze at a light. Make sure you can't see any bright lights at your stargazing site.

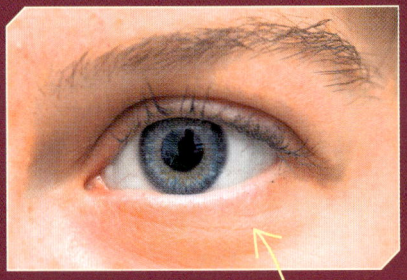 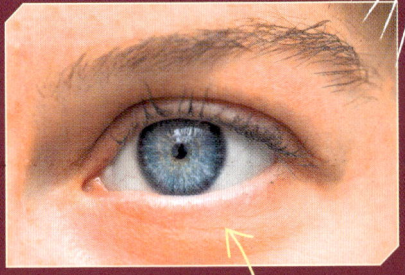

dilated pupil   undilated pupil

 DID YOU KNOW?
Your pupil dilates to about 8 mm wide when you are adapting to the dark.

RED LIGHT

If you want your eyes to stay adapted to the dark, but still see well around you, you can use a red light torch. Red light doesn't spoil your dark-adapted vision. Have a go at making your own red light torch.

HOMEMADE RED TORCH

Attach a red sweet wrapper with a rubber band, or make a cap out of red film to convert your normal torch into a red one.

SEE FAINT OBJECTS

When you're looking at something faint in a telescope, you might notice that it is harder to see when you're looking straight at it. It may even disappear! Don't worry, this is normal. You have a blind spot at night, right at the centre of your vision. Astronomers use averted vision, which means looking slightly to the side of an object. This takes some getting used to. But once you master it you'll be using it all the time!

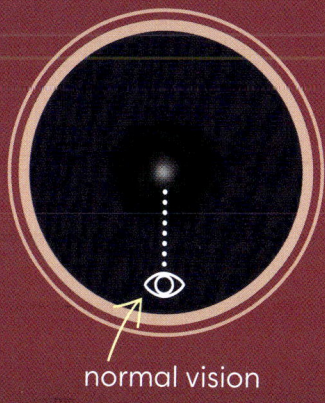

normal vision

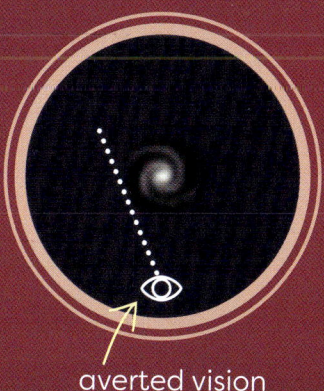

averted vision

BINOCULARS

A pair of binoculars 'supercharges' your eyes by gathering a lot more light for you. Binoculars also magnify (enlarge) an image so that you can see finer details.

EYEPIECE

DIOPTRE

SIZE

'10x50' is a popular size of binoculars for stargazing. They are portable but powerful, and light. '10x' means that things you see will appear ten times bigger. '50' refers to the width of the lenses in millimetres. The wider the lenses, the more light they can gather.

FOCUS WHEEL

LENS

SETTING UP YOUR BINOCULARS

1. Close your eye on the side with the dioptre and focus the binoculars on something in the distance.
2. Then switch so that only the eye on the dioptre side is open.
3. Adjust the dioptre by twisting it until the image appears clear.
4. Now hold up your binoculars to the stars in the night sky.
5. Adjust the focus wheel at the centre of the binoculars to make the image clear.

BINOCULAR VIEWS

Despite what we see in cartoons and TV shows, binoculars do not produce a binocular-shaped image. The view should be a perfect circle when the binoculars are correctly set up.

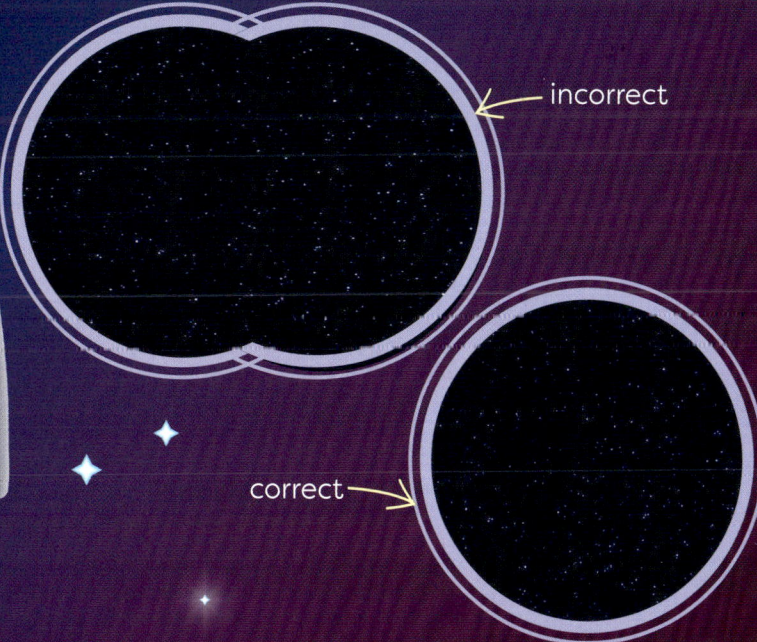

incorrect

correct

TELESCOPES

Telescopes reveal faint or distant things in the sky that you would struggle to see with just your eyes. They can show you details on the Moon, our neighbouring planets and other objects.

EYEPIECE

A telescope gathers light from the sky and brings it to a point of focus, forming an image. An eyepiece is then placed into the telescope to magnify (enlarge) the image.

FILTERS

Filters can be screwed onto the front of the eyepiece. Filters help improve your view of the Solar System by showing better contrast.

Good contrast means that what you are looking at will have very bright highlights alongside very dark shadows, making it easier to pick out features.

 In the Solar System section where you see this symbol, you'll find recommendations for which filters to use.

 DID YOU KNOW?

The telescope was invented by a Dutch lens maker called Hans Lipperhey in 1608. Galileo Galilei and Thomas Harriot were among the first people to build one to use for astronomy.

APERTURE

This is the width of the telescope lens or mirror. A larger aperture gathers more light and allows more magnification.

MOUNT AND TRIPOD

Your telescope will need a mount and tripod to keep it steady.

STAR CHARTS

HOW TO USE STAR CHARTS

Star charts are maps of the night sky. They show the positions of the constellations and deep sky objects. The star charts on the following pages show what you can see in Spring, Summer, Autumn and Winter if you are looking north, south, east and west.

Remember that you can find true north using the stars (see page 28) to help you work out which direction you're facing outside.

The sky is always changing position as the Earth turns in space and moves around the Sun. The charts show the night sky at about midnight at the start of each season. As the season goes by, they show how the sky will look earlier in the evening, for example, in winter around 12 a.m. in December, 10 p.m. in January and 8 p.m. in February.

The sky might not look exactly like the charts as it depends on when and where you are stargazing, but they will help you make a plan of what to look for and where to find it in the sky. The more you become familiar with the locations of the constellations and the deep sky objects, the easier it will be to hop from one to another to find what you are looking for.

Symbols are used to show different deep sky objects:

- ○ Open star clusters
- ⊕ Globular clusters
- ○ Emission nebulae
- ⌖ Planetary nebulae
- ⬭ Galaxies

To learn more about clusters, nebulae and galaxies, see pages 122-139, and for meteor showers, go to pages 116-117.

 On some charts, you will also see this symbol showing the location in the sky where a meteor shower comes from every year. You can find a list of annual meteor showers on page 117.

The ecliptic is marked on each chart. This is the imaginary path that the Sun follows throughout the year (see pages 30-31). The Moon and planets never move too far away from this line, but as their position in the sky is constantly changing, they are not shown on the charts.

The brightest regions of the Milky Way are shown on the charts and look like this.

On star charts, larger dots represent bright stars, and smaller dots represent fainter stars.

49

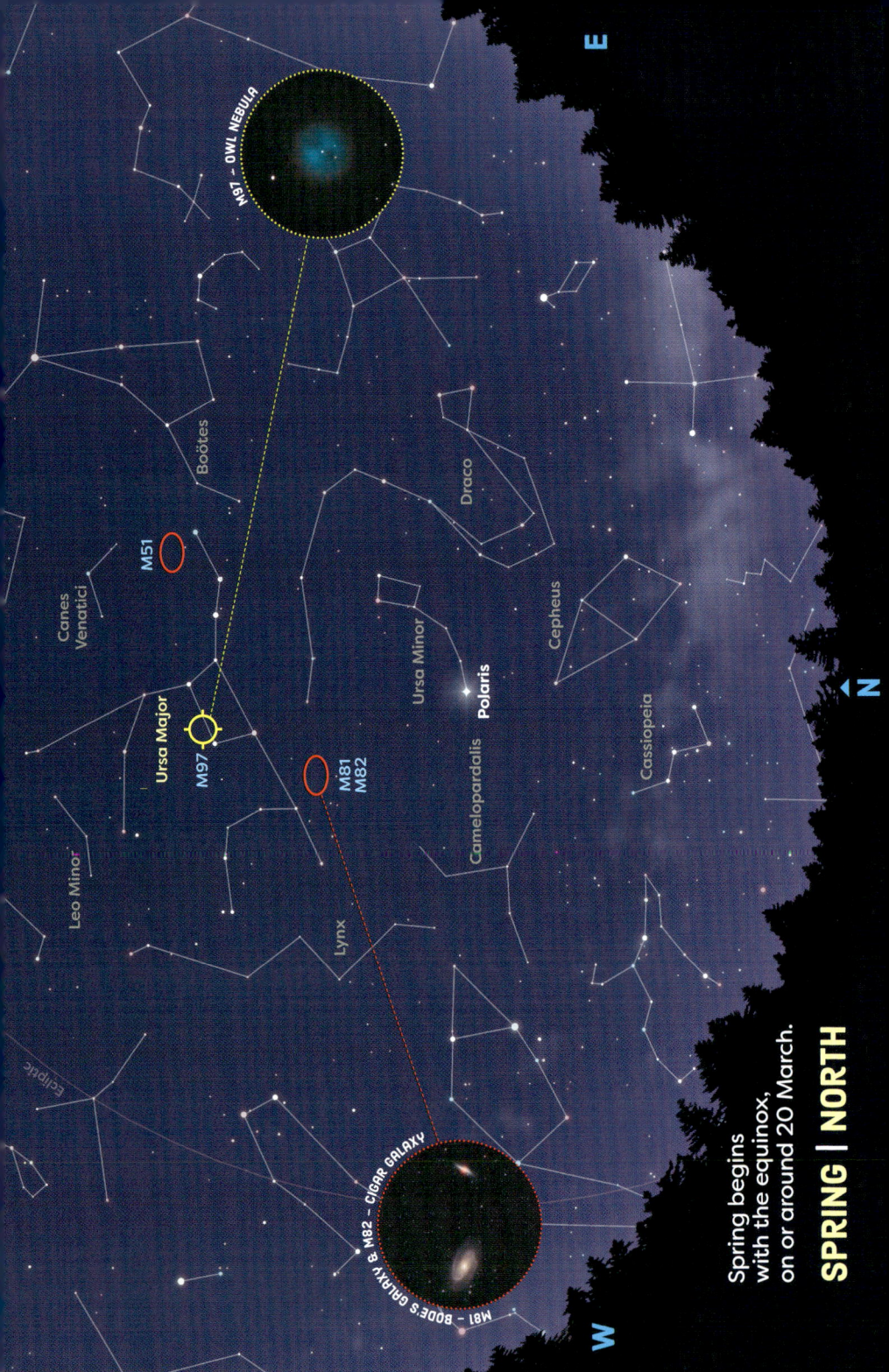

URSA MAJOR

Use binoculars or a telescope to find the galaxies near the great bear's head and tail, as well as the **M97 – Owl Nebula**. Can you spot the owl's eyes?

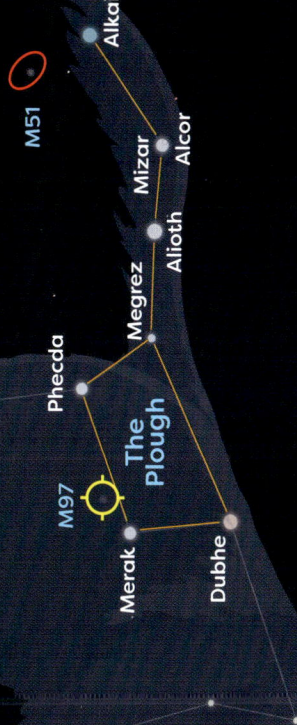

LOOK FOR:

The constellation **Ursa Major**, the great bear, is highest in the sky on spring nights. This is a great time to explore this constellation and find some nearby deep sky objects.

M81 and **M82** are a pair of galaxies to the north of the bear's head. Their common names are **Bode's Galaxy** and the **Cigar Galaxy**. With binoculars or a telescope, you can see them as two patches of light next to each other.

The **M97 – Owl Nebula** is a planetary nebula in Ursa Major. It is just visible in binoculars, but with a telescope you can make out its perfectly circular shape. It has two dark regions that give the impression of large eyes, just like those of an owl.

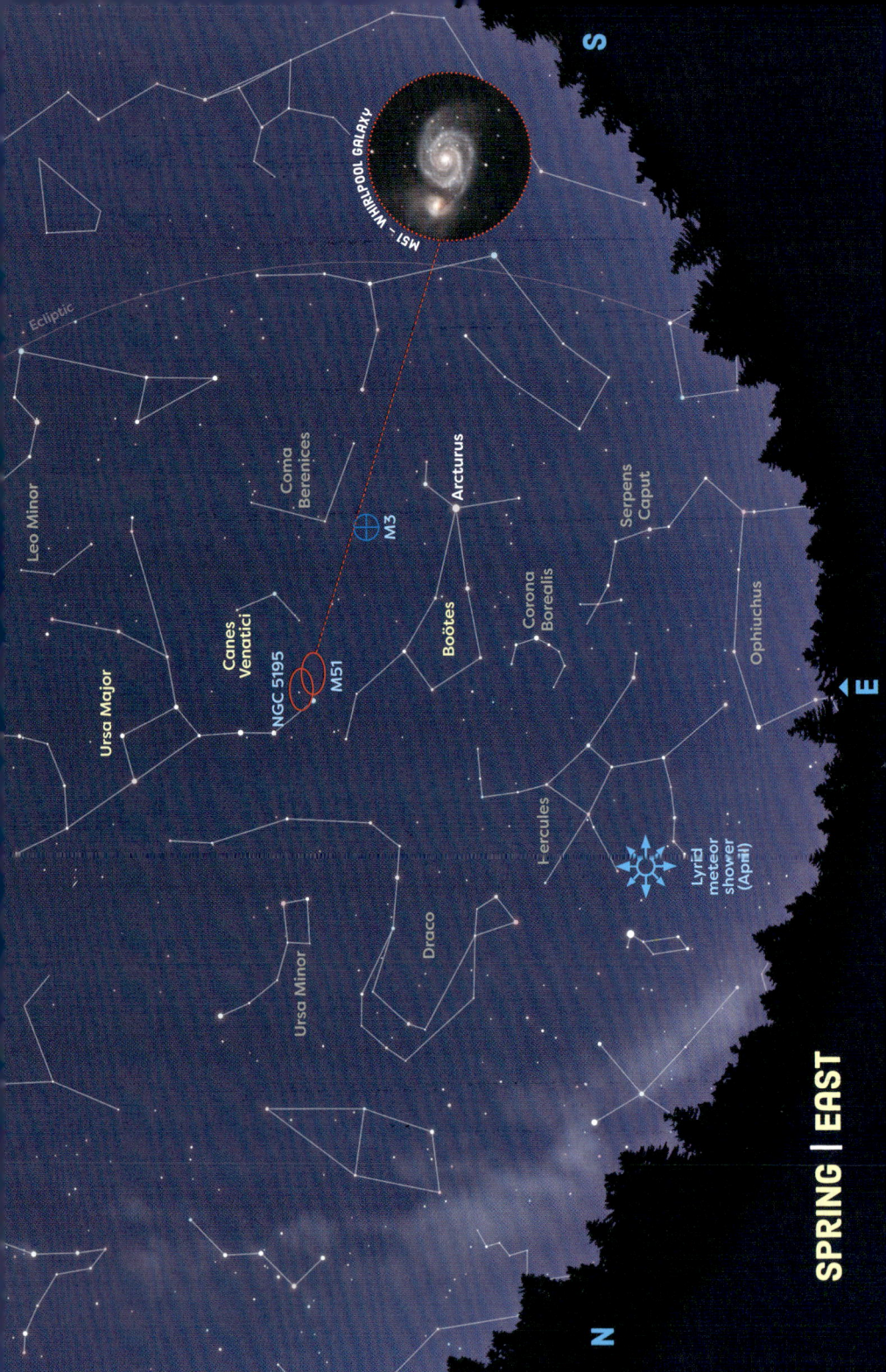

BOÖTES

Boötes follows **Ursa Major** (the great bear) with his hunting dogs **Canes Venatici**. The constellation looks like a kite. Look nearby for the globular cluster **M3**.

LOOK FOR:

Arcturus is one of the brightest stars in the sky. When it rises in the east on late spring evenings it means summer is on the way. Arcturus is a giant star, about 25 times the size of the Sun. Its name means 'guardian of the bear' and it belongs to the constellation **Boötes** the herdsman. Can you see the star's golden orange colour at the bottom of Boötes?

The **M51 – Whirlpool Galaxy** is a very famous galaxy. It is found near the tail of the constellation **Ursa Major**. It is visible in binoculars as a faint patch, but with a telescope, you can see its spiral shape. Next to M51 is a smaller galaxy called **NGC 5195**. M51 is tearing NGC 5195 apart as they drift past each other.

The constellation **Canes Venatici** is the home of the Whirlpool Galaxy. It represents two hunting dogs, one for each star. They have long been considered to be the hunting dogs of **Boötes** the herdsman. The dogs are known as Asterion and Chara.

53

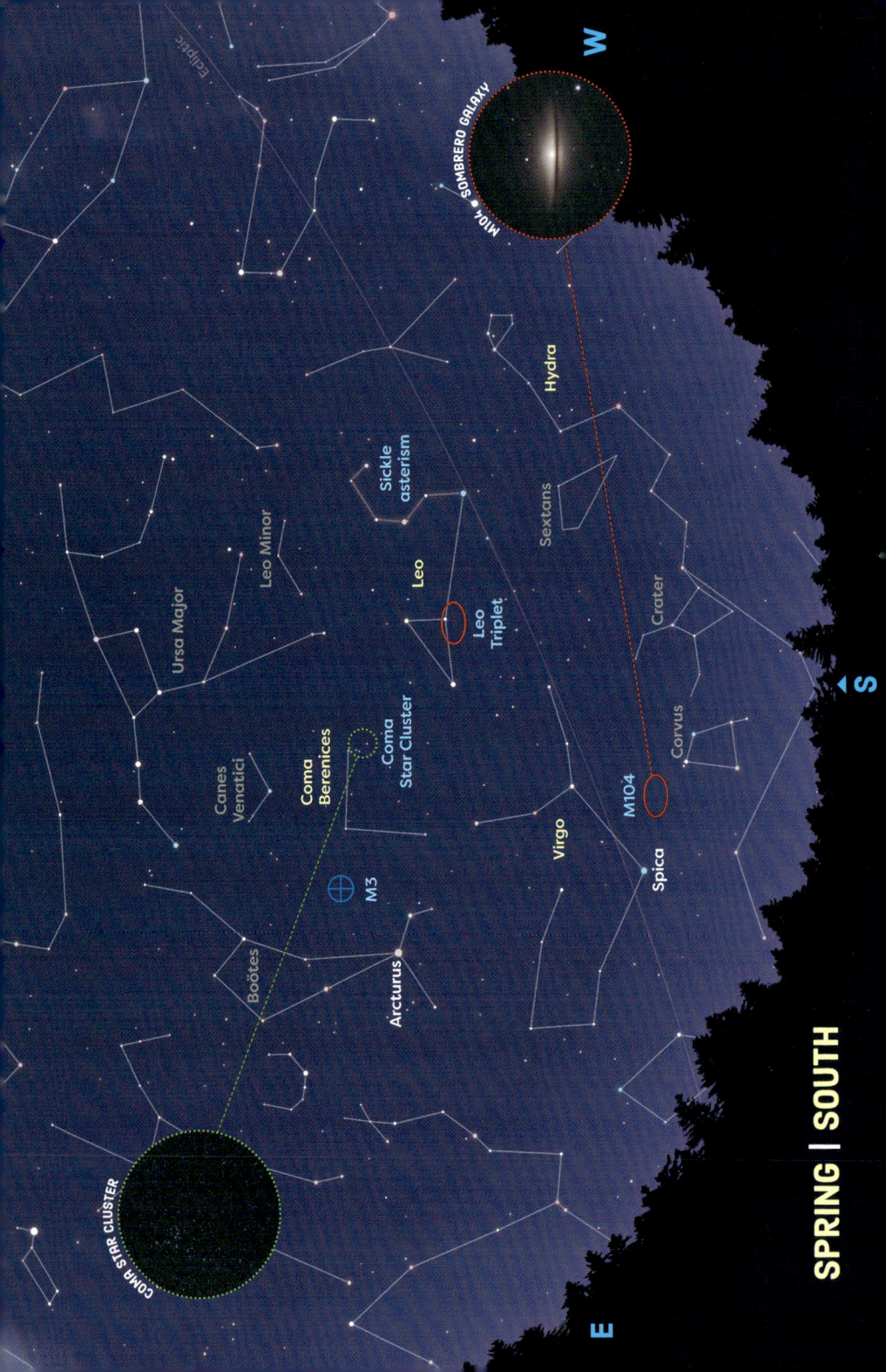

LEO

Spring is galaxy season! There are many galaxies hiding among **Leo.** How many can you find with binoculars or a telescope? Look out for the **Leo Triplet** galaxy.

LOOK FOR:

The **Sickle asterism** is found in the constellation **Leo** (the lion). It looks like a fishing hook or backwards question mark. This asterism makes up the head of the lion.

The **Coma Star Cluster** is a sparkly star cluster. You can spot it by looking to the east of Leo and finding the constellation **Coma Berenices.**

M3, the globular cluster, is a great sight through binoculars or a telescope. It is over 30,000 light years away and can be found northwest of the star **Arcturus**.

Hydra, the water snake, is the largest constellation in the sky, and it stretches along the southern horizon. Can you trace all of its stars from the head in the southwest to the tail in the southeast?

The constellation **Virgo** is home to the famous star **Spica**, which is about 250 light years away. The **M104 – Sombrero Galaxy** lies just west of Spica – the centre of this galaxy is home to a huge black hole.

GEMINI

The twins of the constellation **Gemini** are marked by the stars **Pollux** and **Castor** (below). Castor is a multiple star system made up of 6 stars. Near Castor's feet is the star cluster **M35**.

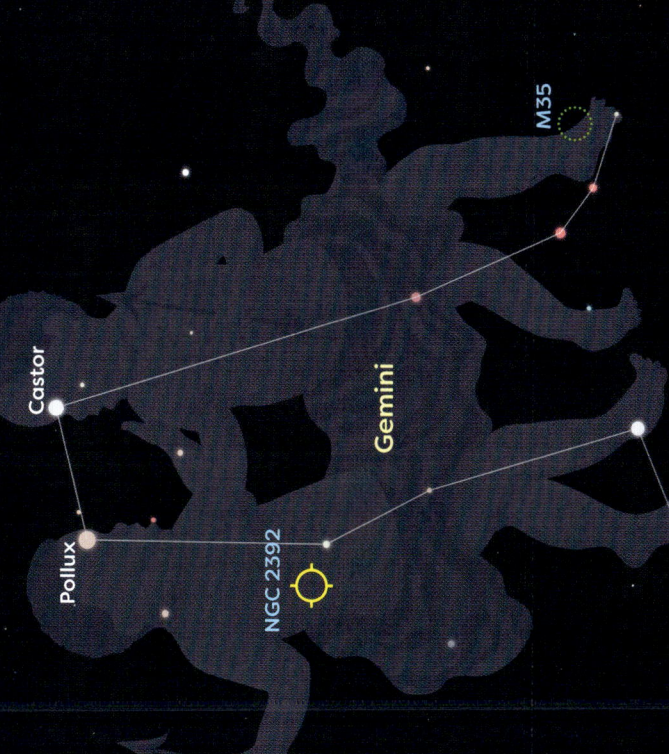

LOOK FOR:

The **M44 – Beehive Cluster** is a gorgeous sight at the heart of the faint constellation **Cancer** the crab. It can just be seen by eye on a very dark and clear night, but it is best viewed through binoculars or a telescope. The M44 – Beehive Cluster is home to around 1,000 stars in total. How many can you count?

The constellation **Lynx** can be found to the south of the constellation **Ursa Major**, the big bear.

Between the faint stars of Lynx and the bright twin stars of the constellation **Gemini** lies a very curious deep sky object, nicknamed the Intergalactic Wanderer. This is a globular cluster labelled **C25** or NGC 2419. Despite being a challenge to see, it is popular with stargazers because it is very far away – about 300,000 light years! This means it drifts in the space between galaxies.

The planetary nebula **NGC 2392** is found in the constellation Gemini and can be seen with a telescope.

57

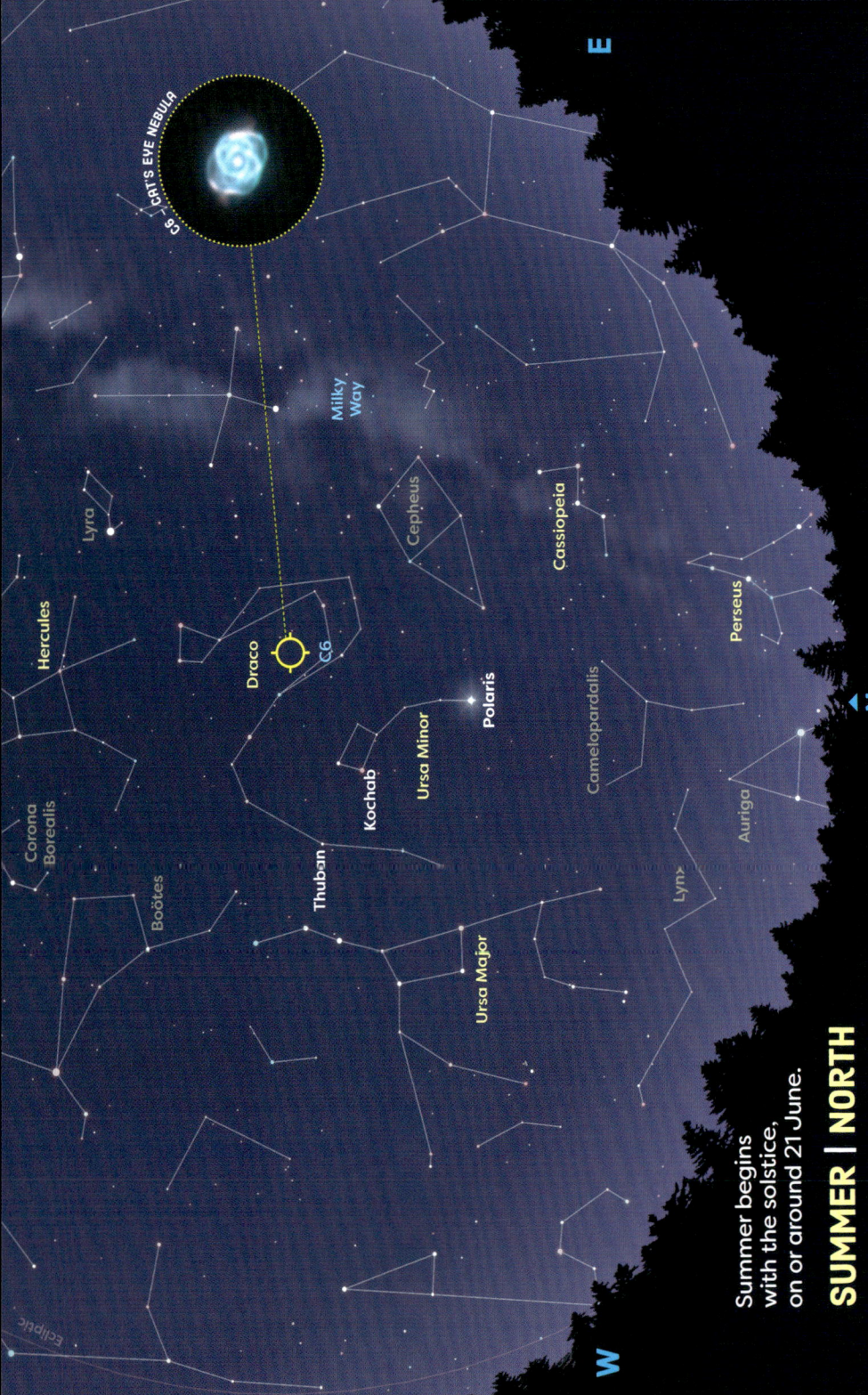

LOOK FOR:

The **C6 – Cat's Eye Nebula** is found to the northwest of the constellation **Hercules**, tucked into the neck of **Draco** the dragon. It is visible in binoculars and telescopes, but it is very small. With a telescope, you might be able to see its ring-like shape. This is one of the most challenging and interesting planetary nebulae to find.

The constellation **Ursa Minor**, the little bear, is turned with its body upward above the star **Polaris** (the North Star). The next brightest star in Ursa Minor is **Kochab**. It has a pale golden-orange colour.

The brightest region of the **Milky Way** is seen in the south, but you can trace it through the sky to the north until you reach the constellations **Cassiopeia** and **Perseus**. There are lots of clusters and nebulae in this area. Explore it with binoculars and see what you find!

DRACO

Draco's long tail reaches between **Ursa Major** and **Ursa Minor** – the two bears.

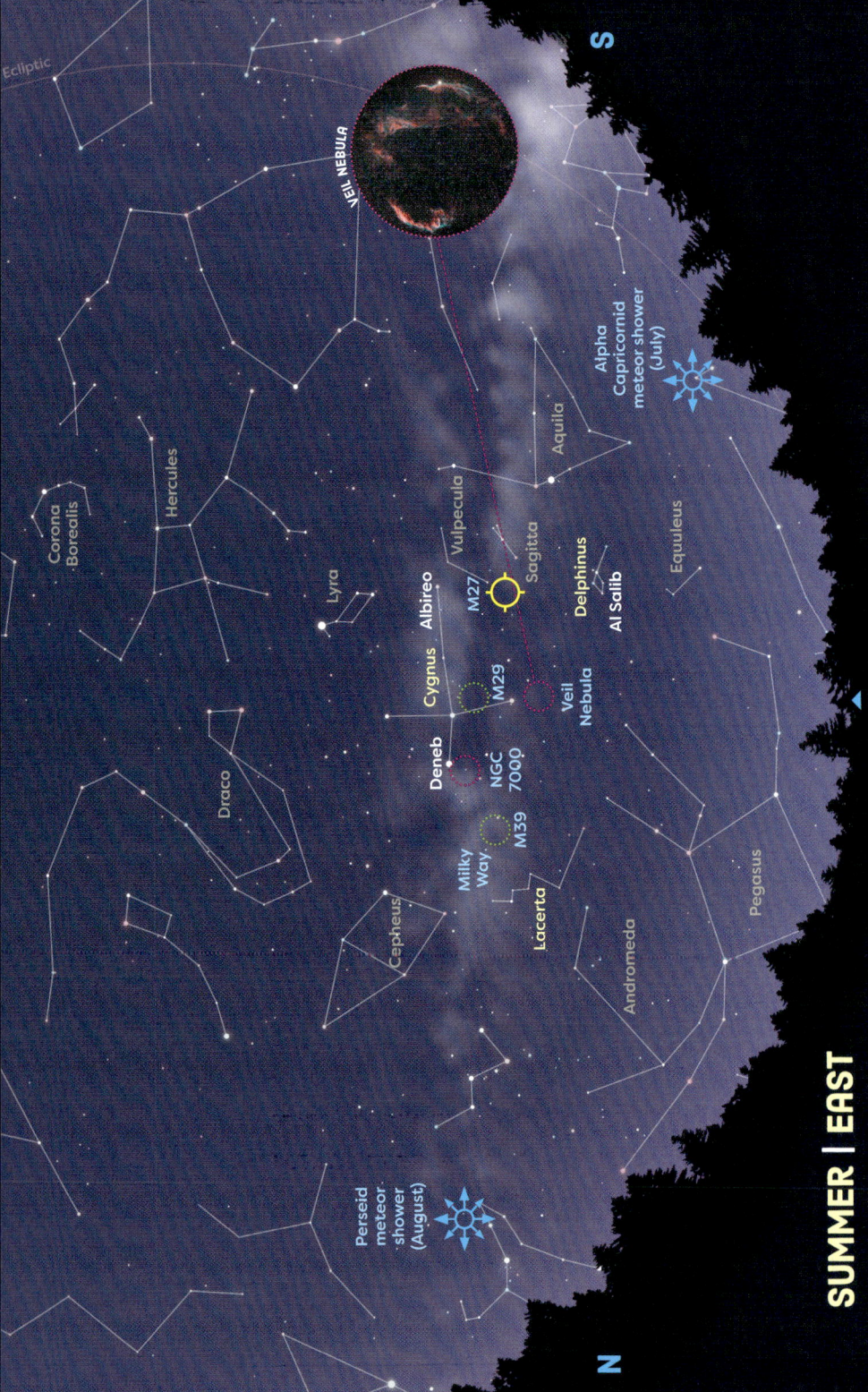

CYGNUS

The star clouds of the constellation **Cygnus** form one of the brightest parts of the **Milky Way**. **Albireo**, the star marking the swan's beak is a beautiful double star – one is blue and one is gold. **Deneb** marks the tail of the swan.

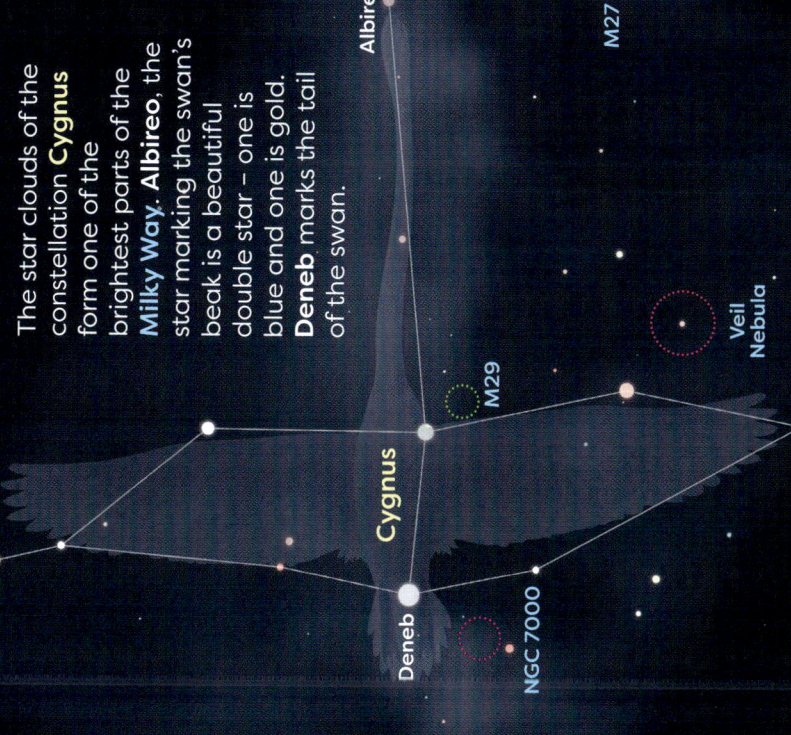

LOOK FOR:

Al Salib is a double star in the nose of the dolphin in the constellation **Delphinus**. Zoom in with a telescope to separate the two stars and see them side by side. The brighter star looks slightly yellow in colour.

The **Veil Nebula**, next to one of the swan's wings in the constellation **Cygnus**, is what remains of a star that exploded over 10,000 years ago. It is an unusual form of nebula called a supernova remnant. The Veil Nebula is very large and is best seen using binoculars.

The constellation **Lacerta** the lizard seems to be crawling towards the **Milky Way**. In front of its nose, there are several beautiful star clusters for you to discover using binoculars.

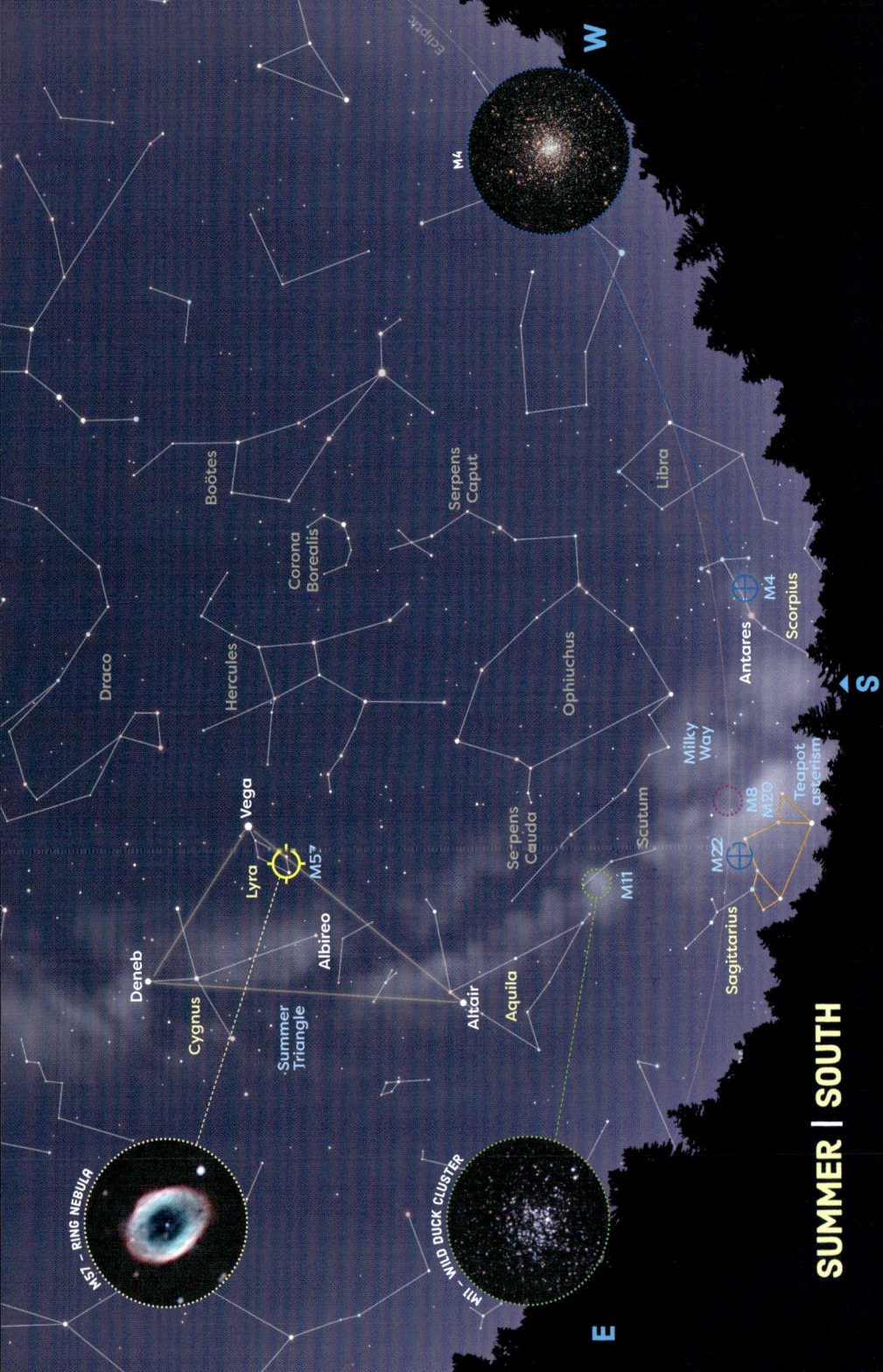

THE TEAPOT & THE MILKY WAY

The **Teapot asterism** in the constellation **Sagittarius** points the way to the centre of the **Milky Way**. There are lots of beautiful star clusters and nebulae here.

LOOK FOR:

The **Summer Triangle** is a large asterism made up of the following three bright stars: **Deneb** (in the constellation **Cygnus** the swan), **Vega** (in the constellation **Lyra** the harp) and **Altair** (in the constellation **Aquila** the eagle). The **Milky Way** runs through the Summer Triangle and down to the southern horizon on summer nights.

The star **Albireo** marks the beak of **Cygnus** the swan. It is close to the centre of the Summer Triangle. Albireo is a double star – a pair of stars that appear close together in the sky. With a telescope, you can see the separate stars: one is golden and the other is blue.

The **M57 – Ring Nebula** hides among the stars of **Lyra** the harp. The string box of the harp is made up of four bright stars. M57 is about halfway between the lower pair of stars. Through a telescope it looks like a little ring of smoke in the sky, which is really the remains of a dying star.

The star **Antares** marks the heart of the constellation **Scorpius** the scorpion. 'Antares' means 'rival of Mars'. Its name comes from its deep orange colour, which reminded stargazers of Mars in the sky.

63

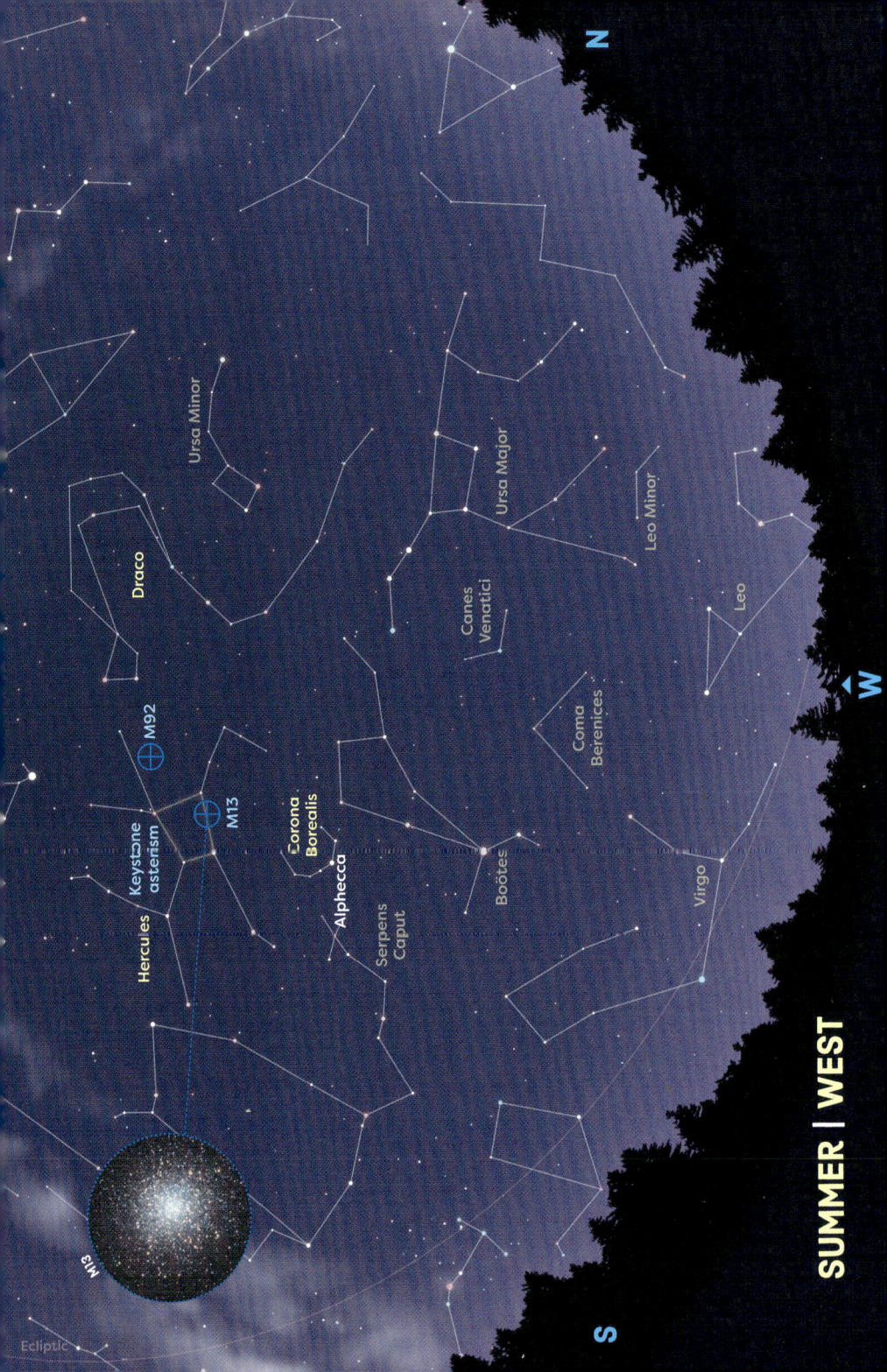

HERCULES

Can you find the lesser known cluster, **M92**, in the constellation **Hercules**?

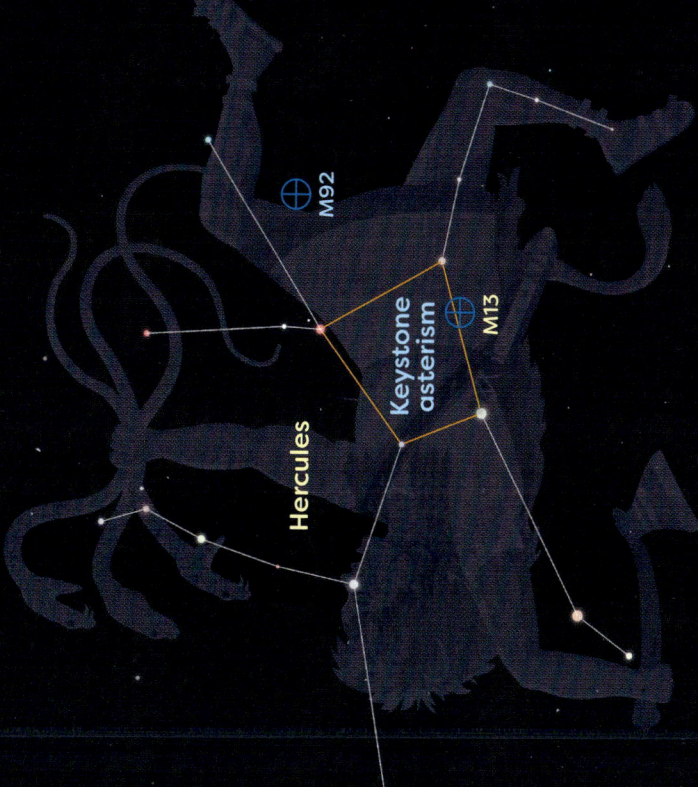

LOOK FOR:

The **Keystone asterism** is made up of the four stars in the centre of the constellation **Hercules**.

The constellation **Draco** the dragon can be found alongside the constellation Hercules, the son of Zeus.

M13 is a great globular cluster that can be found along the western edge of the Keystone asterism. On a dark night, you'll see a very faint patch of light there – this is M13. It is about 25,000 light years away and is home to hundreds of thousands of stars.

Corona Borealis is a small constellation representing a crown or wreath. It is an ancient constellation connected with the story of the Greek god Dionysus, who gave a crown to the Princess of Crete. Its brightest star is **Alphecca**, which is 75 light years away.

PERSEUS

Mirfak is the brightest star in the constellation **Perseus**, the hero. It seems to be surrounded by a wonderful star cluster. Nearby, the **M34 – Spiral Cluster** is another curious sight.

LOOK FOR:

The constellation **Cassiopeia** reaches its highest position at this time of year. Instead of making its usual 'W' shape, it looks like an 'M'. Cassiopeia is a beautiful constellation, worth exploring with binoculars and telescopes. In mythology, Cassiopeia is the queen of Ethiopia and Andromeda's mother.

At this time of year, the constellation **Ursa Major** is low in the north at night.

Camelopardalis is a faint constellation representing a giraffe, an animal once likened to a 'camel-leopard' because of its markings.

Kemble's Cascade is a chain of stars in the constellation Camelopardalis. More than 20 stars seem to form a nearly straight line, which is visible in binoculars and telescopes. The stars are not as close together as they appear, so Kemble's Cascade is an asterism rather than a star cluster.

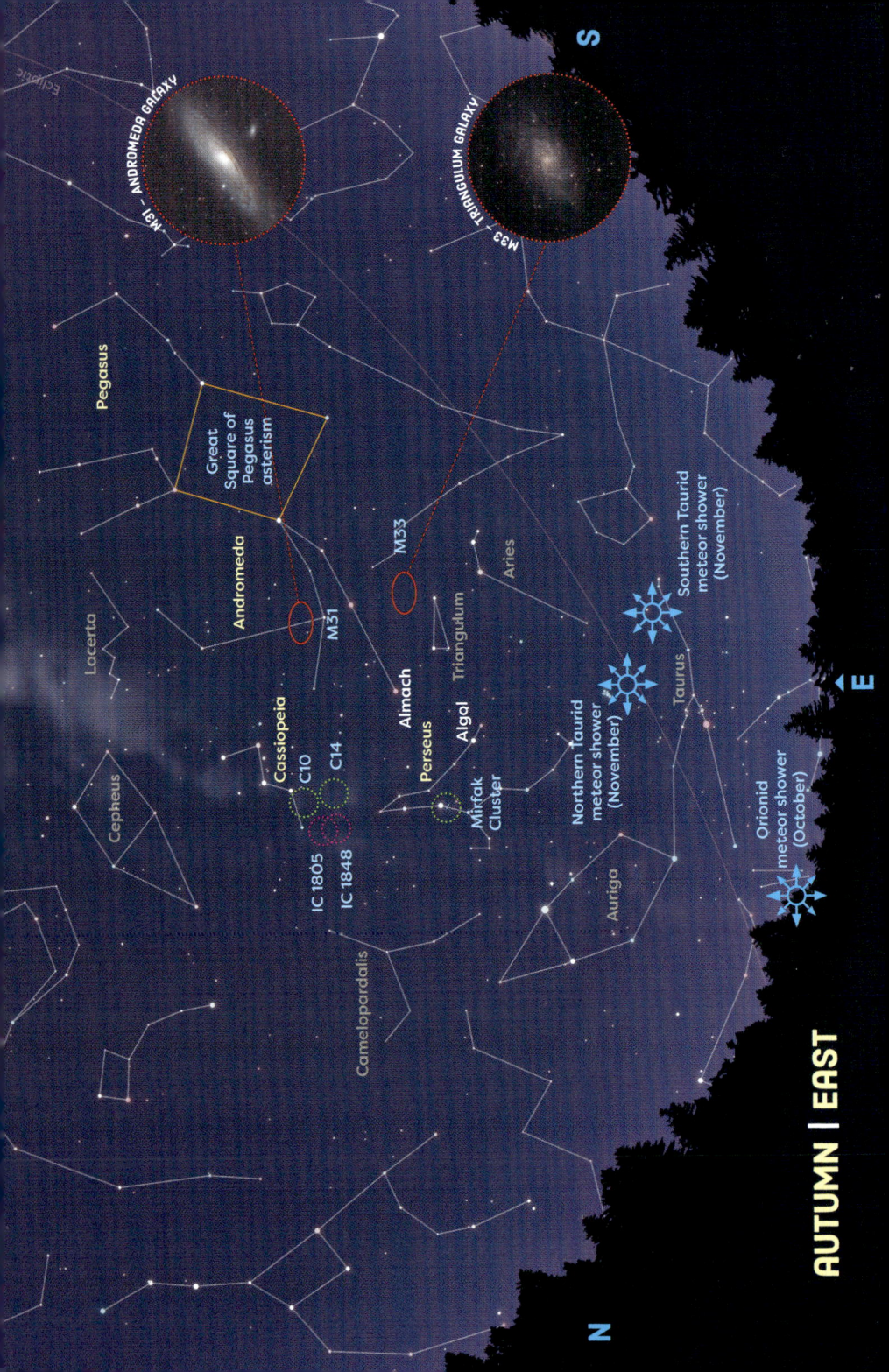

Cassiopeia

CASSIOPEIA

Cassiopeia is home to a few star clusters. It is also close to the famous **C14 – Double Cluster**, where two star clusters appear close together in binoculars or a telescope. Nearby, look for two nebulae – the **IC 1805 – Heart Nebula** and the **IC 1848 – Soul Nebula**.

Double Cluster

C10

Heart nebula

Soul nebula

LOOK FOR:

The constellation **Andromeda** is connected to the **Great Square of Pegasus** asterism. Andromeda is famous for being the home of the Milky Way's nearest neighbouring galaxy, the **M31 – Andromeda Galaxy**. You can see it with just your eyes on a dark night, and it is the most distant thing in the Universe visible without binoculars or a telescope.

The **M33 – Triangulum Galaxy** is a smaller neighbour of both the Milky Way and the **M31 – Andromeda Galaxy**. It can be found using binoculars.

Almach is a yellow-blue double star in Andromeda. See the separate stars with a telescope.

Perseus is the legendary hero who rode Pegasus and saved the princess Andromeda. In its heart is the star **Mirfak**, which is surrounded by a star cluster and is a lovely sight in binoculars. Perseus is also home to **Algol**, the Demon Star.

AQUARIUS

Aquarius is home to two globular clusters that can be seen with binoculars. Look close to **M72** and find a small, Y-shaped pattern of stars called **M73**. This looks like a tiny star cluster, but it is really an asterism of stars that appear to be closer than they are.

LOOK FOR:

The **Great Square of Pegasus** is an asterism. It is made up of four stars in the body of the constellation **Pegasus**. Pegasus is the legendary flying horse belonging to the hero Perseus. How many stars can you count inside the Great Square of Pegasus?

The **Circlet asterism** can be found beneath the Great Scuare of Pegasus. It is a ring of five stars representing one of the fishes in the constellation **Pisces**.

The **C22 – Blue Snowball Nebula** can be found above the Great Square of Pegasus, close to the constellation **Lacerta** the lizard. With a telescope, you can see a point of light surrounded by a fuzzy circle. The point is a dying star and the fuzzy circle is its atmosphere escaping into space. The nebula takes its name from its blue colour, which can be seen in photographs.

71

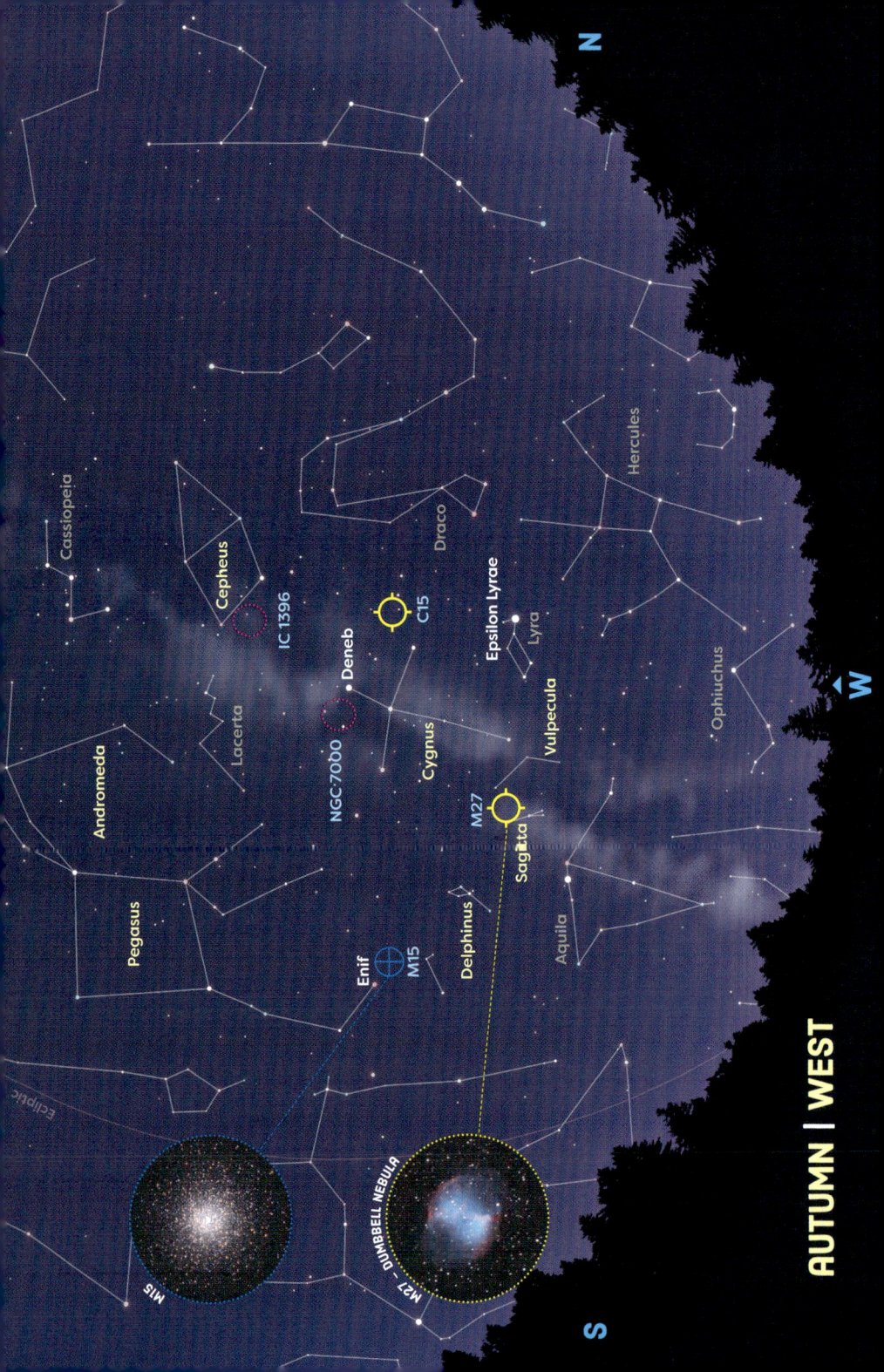

LOOK FOR:

The Cygnus star clouds glow around the star **Deneb** at the tail of the constellation **Cygnus**.

The constellation **Cepheus** is found above the constellation Cygnus. Cepheus is home to several star clusters and nebulae, and it looks very interesting in binoculars. Look near the western and southern edges of Cepheus to find the **IC 1396 – Elephant's Trunk Nebula**. In mythology, Cepheus is the King of Ethiopia and is father of the princess **Andromeda**, who appears as a nearby constellation.

M15, the globular cluster (also known as the Pegasus Cluster) is to the east of the small constellation **Delphinus**. It is one of the most famous and popular globular clusters. It is very close to the star **Enif**, in the nose of **Pegasus**.

VULPECULA AND SAGITTA

Vulpecula and **Sagitta** are small constellations. Look for the globular cluster **M71**, the **M27 – Dumbbell Nebula** and a cluster-like asterism called the **Coathanger**. You will need binoculars or a telescope to see them.

URSA MINOR

Polaris is the Pole Star, or North Star. It sits very close to the north pole, and points the way to true north all night.

LOOK FOR:

The constellations **Cassiopeia** and **Ursa Major** lie to the left and right side of the star **Polaris** (the North Star).

Ursa Minor, the constellation of the little bear, stands upright with its tail curving upwards.

The **C1 – Polarissima Cluster** can be found to the left of Ursa Minor's tail. It is the nearest star cluster to Polaris and is very old and distant. It is challenging to see it, so use binoculars or a telescope on a dark night.

The constellation **Draco** the dragon snakes down towards the horizon, with its tail pointing up into the sky. The star at the tip of the tail is called **Giausar**.

Winter is the perfect time to capture the rotation of the Earth in a photograph. Take a long timelapse of the sky with your camera pointing to the north (see page 28). When you play it back, you will see the constellations revolving around Polaris.

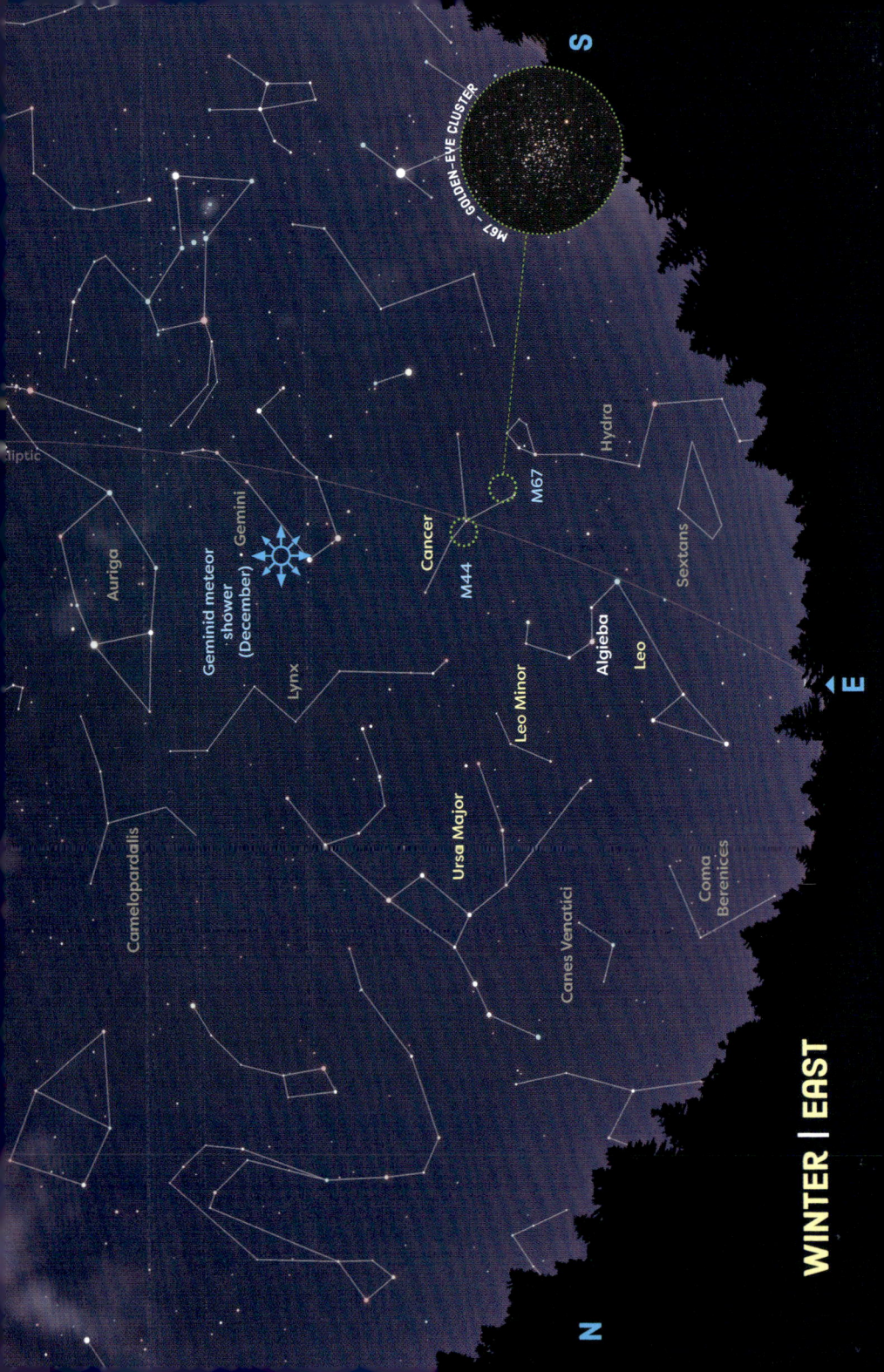

CANCER

Cancer is a faint constellation, but it contains two star clusters. The **M44 – Beehive Cluster** is easy to pick out with binoculars. Can you find the less well-known **M67 – Golden-Eye Cluster** to the south?

LOOK FOR:

The **M67, Golden-Eye Cluster** in the constellation **Cancer** the crab is a small group of at least 500 stars, but you can probably see between 50 and 100 stars through a good telescope. Look to the south of the more famous **M44 – Beehive Cluster** to find it.

Leo Minor is a small constellation between **Ursa Major** and **Leo**. Its name means 'little lion'. Over the long history of stargazing, people disagreed on whether to include the stars in Leo Minor as part of another constellation. The constellation has no bright stars, making it a challenge to find.

Algieba is a double star found in the constellation Leo. Use a telescope to separate the two stars. Do you see any colours?

ORION

The bright orange star, **Betelgeuse**, is a red supergiant that will one day explode.

Orion

Orion's Belt
Mintaka
Alnilam
Alnitak
M42
Rigel

Betelgeuse

LOOK FOR:

The **Winter Hexagon asterism** in the southern sky contains six stars. The southernmost star is **Sirius**, which is the brightest star in the night sky. **Capella**, the 'Goat Star' is at the northernmost point.

Orion is home to the stars **Alnitak, Alnilam** and **Mintaka** – also known as **Orion's Belt**. The **M42 – Orion Nebula** is at the centre of Orion's Sword. In photos, it appears pink and blue, but with a telescope, you may also see a hint of green.

Gemini features twin stars **Castor** and **Pollux**. Pollux is a golden yellow star – it's the nearest giant star to Earth. Castor is a multiple star system.

The **Milky Way** runs through the Winter Hexagon, between stars Capella and Sirius. The Milky Way appears fainter in the winter than in the summer. Scan the sky between the constellations of **Auriga** and **Gemini** with binoculars or a telescope to discover a chain of star clusters.

The **M45 – Pleiades Cluster** in the constellation **Taurus** is easily seen by eye and looks particularly brilliant through binoculars.

79

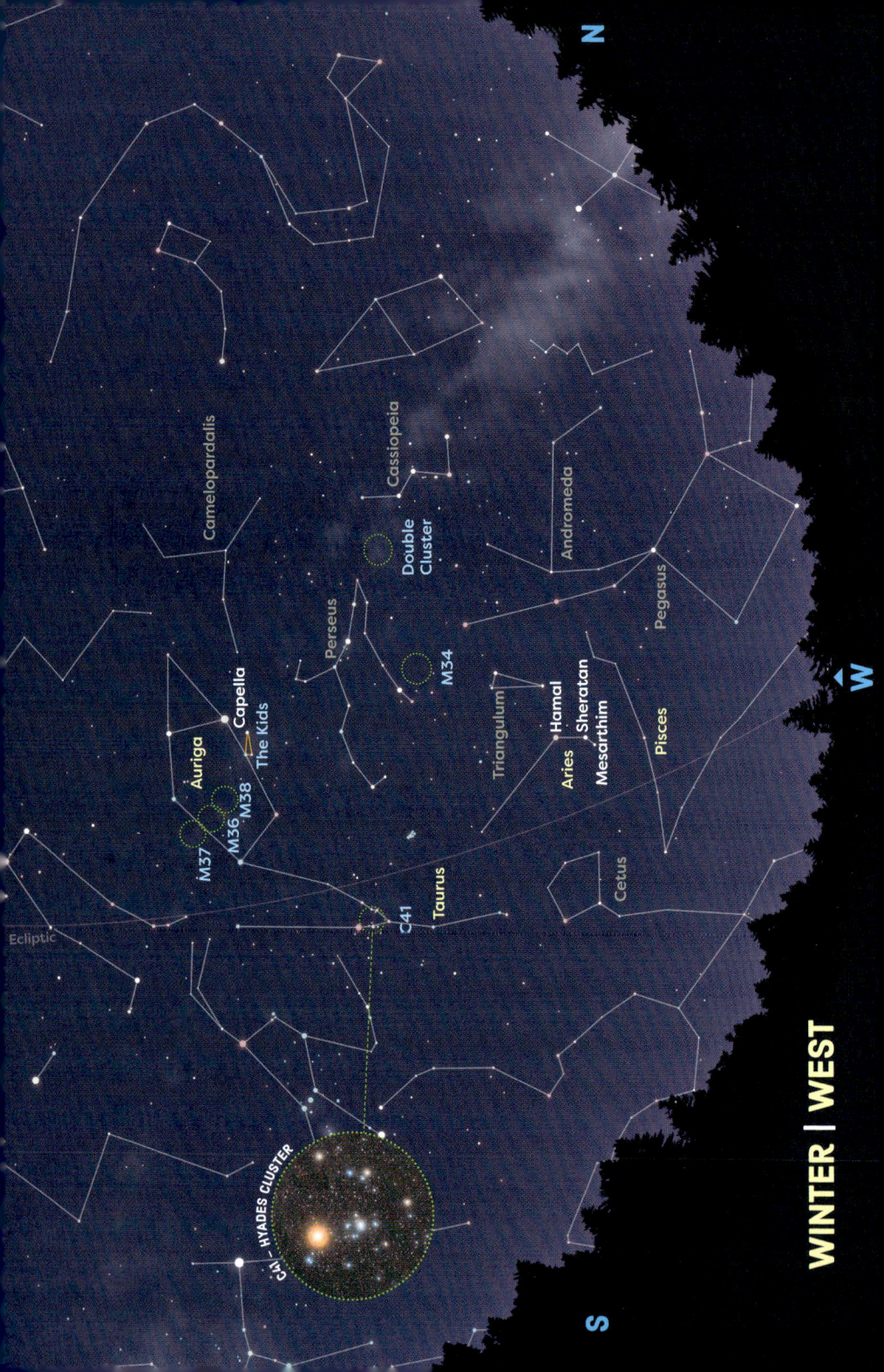

AURIGA

Auriga's brightest star, **Capella**, is sometimes called the Goat Star. Can you find the star clusters?

LOOK FOR:

The star **Capella** is high in the west along with its constellation **Auriga**. In some stories, Auriga is a chariot driver; in others, he is a goat herder. Auriga is a large ring of five stars. Within this ring are several spectacular star clusters for you to find with binoculars or a telescope.

The Kids are a trio of stars next to the star **Capella**.

Aries, the ram, is a zodiacal constellation. It follows **Pisces**, the two fish, down towards the western horizon. The two brightest stars in Aries are **Hamal** and **Sheratan**. Another star of Aries, **Mesarthim**, is particularly interesting. It is a very close double star and you will need to use a high magnification with a telescope to separate it. This is a great challenge for confident telescope users!

The **C41 – Hyades Cluster** is a V-shaped star cluster found in the constellation **Taurus**.

SOUTHERN STARS

The star charts in this book show you the night sky throughout the year in the northern hemisphere. In the southern hemisphere, different constellations and deep sky objects can be seen.

This chart shows the southern skies. It contains constellations that are rarely or never seen from the northern hemisphere. If you have the opportunity to travel to the southern hemisphere, take some time to look at the wonderful constellations that are visible in the skies there.

Which of the southern constellations would you be most excited to see?

Crux (Southern Cross)

Crux is the smallest constellation in the night sky. **Hydra** is the biggest.

THE SOLAR

Asteroid Belt

Sun

Mercury

Venu

SYSTEM

Saturn

Jupiter

Uranus

Mars

Neptune

Earth

THE SPECTACULAR SOLAR SYSTEM

Starting with just your eyes, or by using binoculars or a telescope, you can make a journey of discovery into the Solar System! You can see storm systems on other planets, watch comets sail across the sky or count the craters on the Moon – there's so much to explore.

WHAT IS THE SOLAR SYSTEM?

The Solar System is made up of the Sun and everything that orbits (travels around) it. This includes eight planets and their moons, as well as asteroids, comets and dwarf planets.

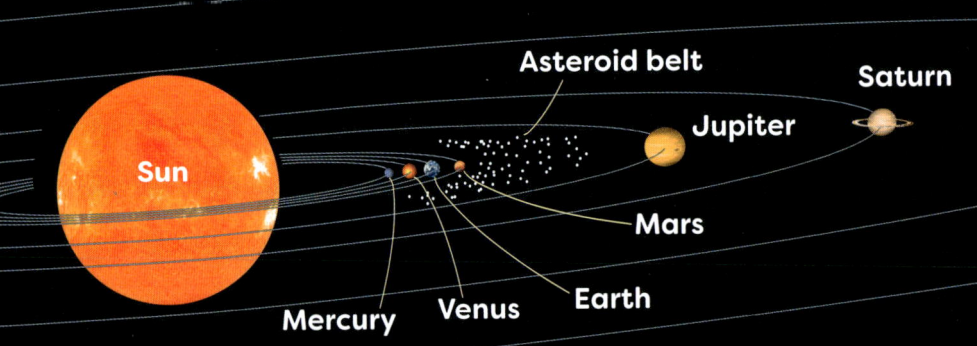

Diagram not to scale

SMALL BUT CLOSE

Compared to stars, nebulae and other galaxies, the Solar System is practically on our doorstep! The planets aren't very far away, cosmically speaking, but they are much smaller than stars.

Mercury, Venus, Mars, Jupiter and Saturn are the five brightest planets in our night sky. They are visible without a telescope. As the planets are always on the move, you can use online resources to help you figure out when and where they are going to appear (see page 162).

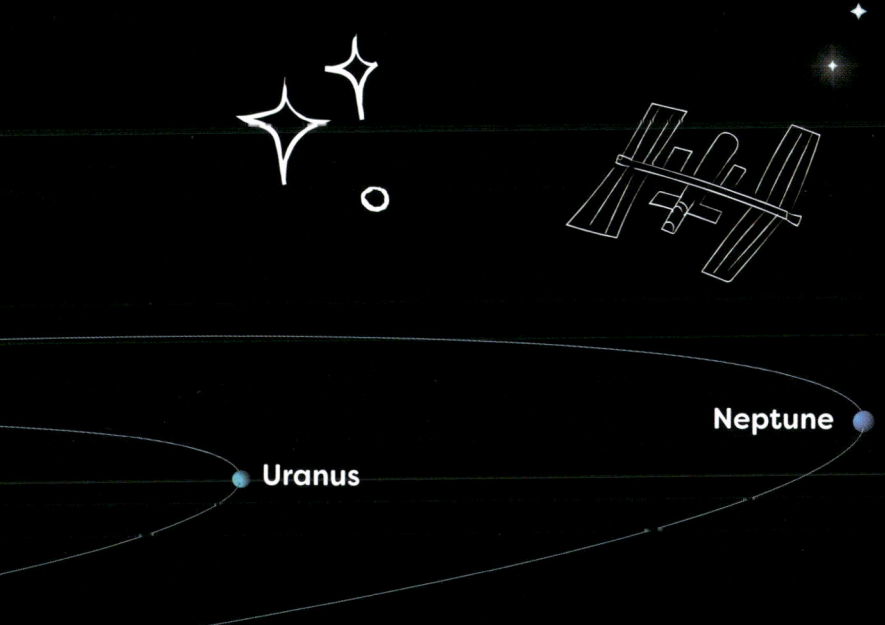

The paths the planets follow around the Sun all share the same plane, meaning they are all on the same level. Imagine peas rolling around on a flat plate, with the Sun in the centre of the plate!

OUR MAGNIFICENT MOON

The Moon is our neighbour in space and a companion to the Earth. It's the brightest object in the sky after the Sun, and its surface has lots of interesting features.

MOONLIGHT

The Moon does not make its own light – moonlight is actually reflected sunlight. Nevertheless, it can seem incredibly bright in the night sky. Moonlight makes stars and other fainter objects harder to see.

DID YOU KNOW?

Other planets also have moons – astronomers sometimes call them 'natural satellites'.

THE TERMINATOR

When you're trying to spot features on the Moon, look closely at the line where day (light side) meets night (dark side). It's called the terminator. Shadows along this line clearly reveal the rugged or uneven surface of the Moon.

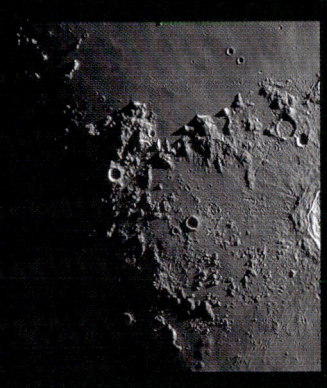

THE MOON WITH BINOCULARS

You can make out many large features on the Moon with binoculars, such as the lunar maria, which are large dark patches caused by ancient volcanic flows. You can also see craters that were formed when huge pieces of rock crashed into the Moon in the distant past.

THE MOON THROUGH A TELESCOPE

A telescope will show you thousands of craters on the Moon's surface and many other features. With a telescope and in very good conditions, you can see features on the Moon that are about two kilometres wide.

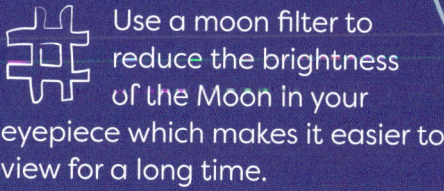 Use a moon filter to reduce the brightness of the Moon in your eyepiece which makes it easier to view for a long time.

A yellow filter (named #8) will improve contrast in small telescopes. Orange (#21) and blue (#80A) filters show even more contrast, but they should only be used in larger telescopes.

10 THINGS TO FIND ON THE MOON

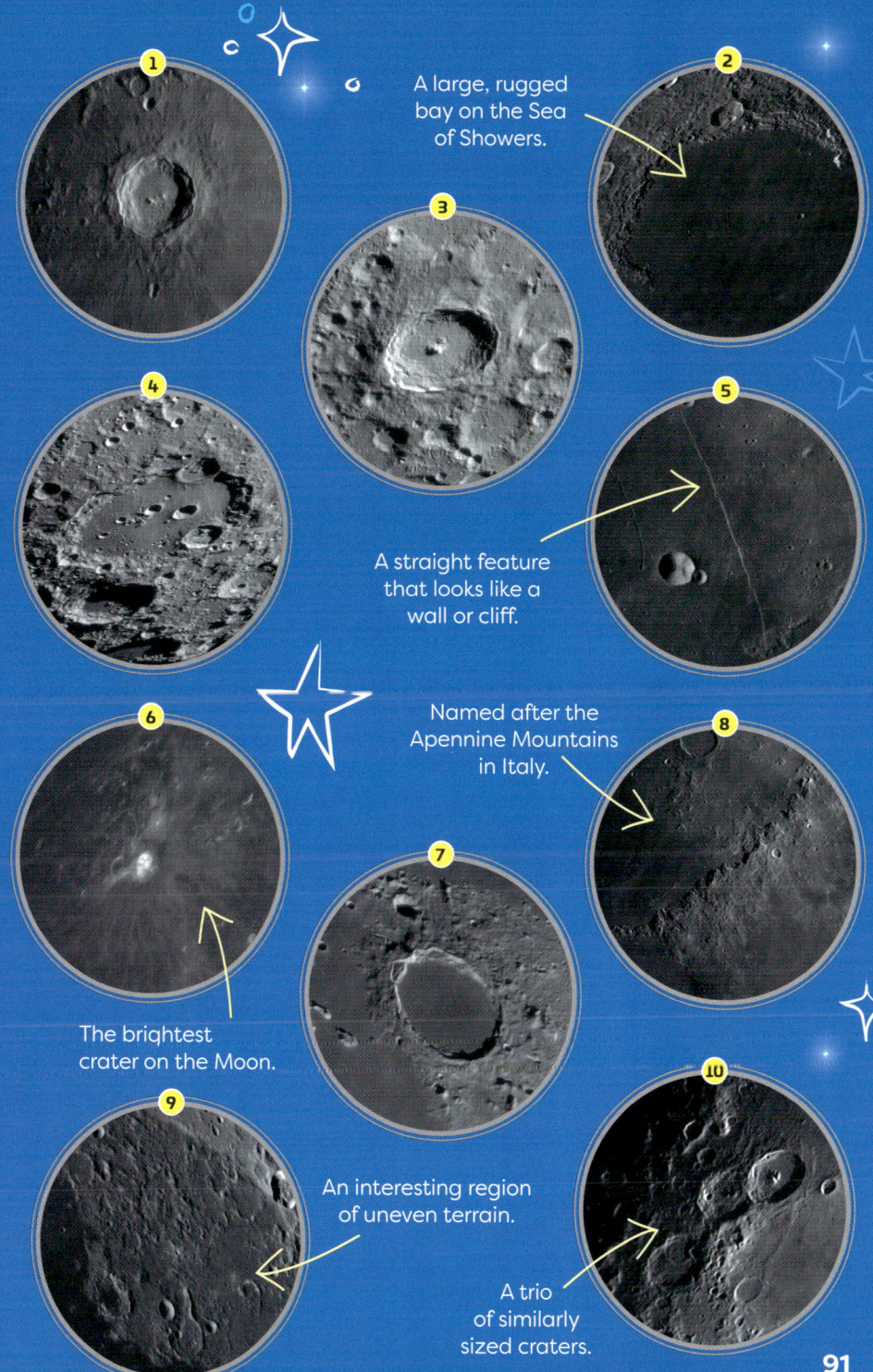

THE MOON'S ORBIT

The Moon orbits (moves around) the Earth once every 27.3 days. It also rotates on its own axis during that time, so we always see the same side of the Moon from the Earth.

Wherever the Moon is, one half of the Moon is always lit by the Sun (day side) and the other side stays in darkness (night side). As the Moon orbits the Earth, we see different amounts of the the Moon's day and night side, depending on its position. This means that sometimes the Moon might look like a narrow crescent and other nights it might look like a bright circle. The different shapes of the Moon that we see at different times of the month are called the moon's phases.

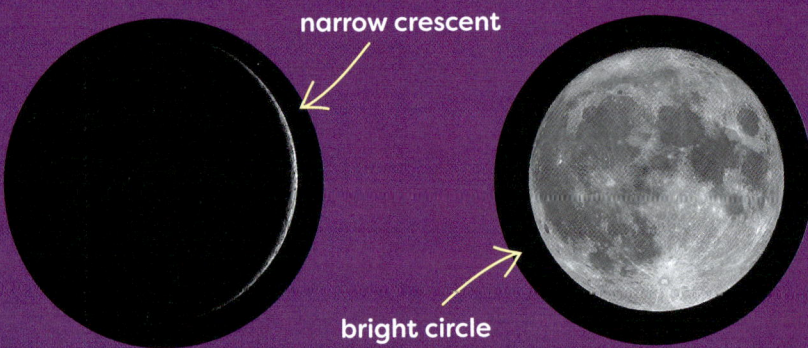

narrow crescent

bright circle

The Moon's orbit isn't a perfect circle (it's elliptical) so the Moon's distance from the Earth changes.

 DID YOU KNOW?

One side of the Moon faces the Earth at all times, and there is a far side of the Moon we never see. Only a small number of astronauts have seen it with their own eyes.

When the Sun and the Moon are in the same part of the sky, the phase is called 'new moon'. Only the night side of the Moon faces Earth, so the Moon is not visible to us.

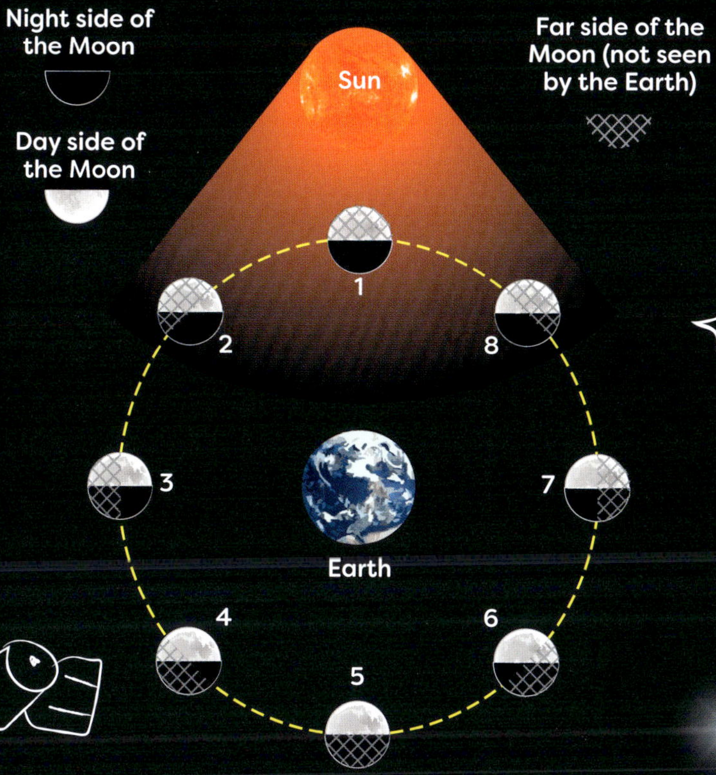

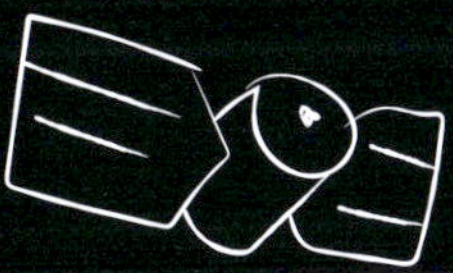

Night side of the Moon

Day side of the Moon

Sun

Far side of the Moon (not seen by the Earth)

1
2
3
4
5
6
7
8

Earth

When the Moon is on the opposite side of the Earth compared to the Sun, we only see its day side. This phase is called 'full moon'.

THE MOON'S PHASES

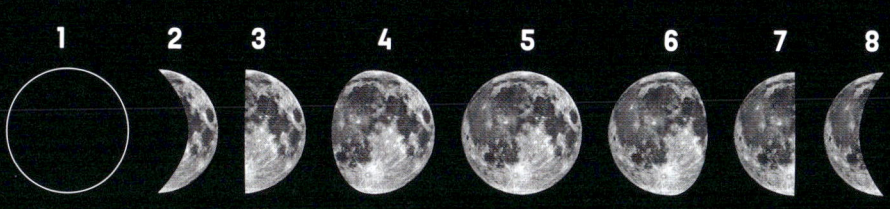

1	2	3	4	5	6	7	8
New	Waxing crescent	First quarter	Waxing gibbous	Full	Waning gibbous	Last quarter	Waning crescent

LUNAR ECLIPSES

On rare occasions, the Moon passes into the Earth's shadow, producing a lunar eclipse. It's perfectly safe – and very special – to watch!

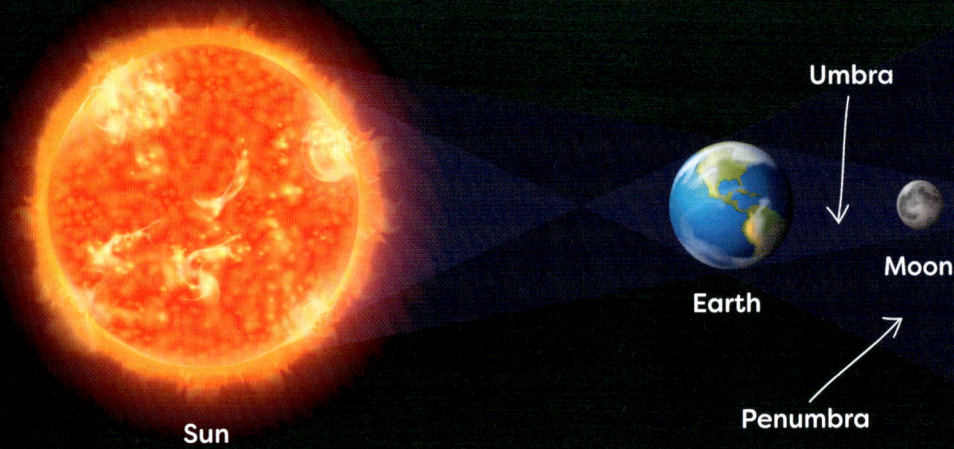

The Earth's shadow has two regions – the penumbra and the umbra. The penumbra is the outer area of the Earth's shadow and the umbra is the very dark central area.

HOW TO SEE A LUNAR ECLIPSE

Lunar eclipses occur on specific dates and are only visible in certain locations. You can use the links in the **Resources** section to discover when the next lunar eclipse will be visible where you are.

TYPES OF LUNAR ECLIPSE

PENUMBRAL LUNAR ECLIPSE

When a part of the Moon enters the outer region of the Earth's shadow – the penumbra – it becomes slightly dimmer. This can be quite hard to notice.

PARTIAL LUNAR ECLIPSE

A partial eclipse happens when a part of the Moon enters the much darker central region of the Earth's shadow – the umbra. When this happens, it is possible to see part of the Moon turn a deep red-orange.

TOTAL LUNAR ECLIPSE

The most dramatic eclipses happen when the Moon is completely within the Earth's umbral shadow. The whole of the Moon takes on a red-orange colour. It's a strange but beautiful sight.

WHY DOES THE MOON LOOK RED DURING A LUNAR ECLIPSE?

Sunlight is usually reflected off the Moon but during a lunar eclipse it has to travel through the Earth's atmosphere first. The white light from the Sun is made up of every colour. When sunlight travels through the Earth's atmosphere, the blue light is scattered, or directed away, so all that's left to travel to the Moon is reddish and orange light.

SOLAR ECLIPSES

Our Sun is incredible. It heats and lights our world, and it provides the energy in our food. It's the heart of our Solar System, and it keeps all the planets in neat orbits.

When the Moon passes between the Sun and the Earth, it is called a solar eclipse. Sometimes the Moon only blocks part of the Sun's light. This is called a partial solar eclipse. Other times, the Moon blocks all of the Sun's light. This is called a total solar eclipse. You should never look directly at the Sun.

SOLAR ECLIPSE

Sun

Moon

Earth

TYPES OF SOLAR ECLIPSE

PARTIAL SOLAR ECLIPSE

The most common and widely seen type is a partial solar eclipse. From the ground, it looks like a bite has been taken out of the Sun by the Moon.

CENTRAL SOLAR ECLIPSE

When the Moon passes directly in front of the Sun, a central eclipse occurs. Three things can happen:

ANNULAR SOLAR ECLIPSE

If the Moon is quite far from the Earth in its orbit, it cannot completely cover the Sun. A ring of light surrounds it. This is known as a 'ring of fire'.

TOTAL SOLAR ECLIPSE

If the Moon is quite close to the Earth in its orbit, it can completely cover the Sun, resulting in a total eclipse. Everything around you will take on a strange darkness for a few minutes. When the eclipse ends, sunlight returns immediately.

HYBRID SOLAR ECLIPSE

On very rare occasions, the Moon's distance from Earth will change enough during the eclipse, so that an annular eclipse becomes a total eclipse or vice versa. This is called a hybrid solar eclipse.

HOW TO SEE A SOLAR ECLIPSE

Partial solar eclipses are visible quite regularly from most places around the world. You can use the links in the Resources section to find when the next one will be. Total solar eclipses are rare and only visible from a small area on Earth so to see one, you will probably need to travel to another country.

WARNING!

Do not look directly at the Sun or point binoculars or a telescope at it. You can get special solar filters for binoculars and telescopes, but always ask a grown-up to help you when observing the Sun.

VIEW THE SUN EASILY AND SAFELY

The safest way to view the Sun is by projecting its image using a pinhole projector.

YOU WILL NEED:

- Two pieces of thick white paper or card
- A needle or drawing pin

WHAT TO DO:

1. Use the needle or drawing pin to make a small hole in the centre of one piece of card.
2. Standing with your back to the Sun, hold up the piece of card so that the Sun shines on it.
3. Hold the other piece of card (without a hole) behind. You will see an inverted image of the Sun projected on the card through the pinhole.

MINIATURE MERCURY

This planet, which is closest to the Sun, is also the Solar System's smallest planet! Little Mercury graces our morning and evening skies but can be tricky to spot! Mercury is a rocky planet, but unlike the Earth, it has no real atmosphere and an unusually large iron core at its centre.

FACTFILE

Diameter: 4,880 km (roughly 3 times smaller than Earth)
Orbital period: 88 days (0.2 years)
Length of a solar day: 176 Earth days
Average distance from the Sun: 57.9 million kilometres
Number of moons: 0
Strength of gravity: 38% of Earth's gravity

LOOK FOR MERCURY AT SUNSET

You can see Mercury with just your eyes shortly after sunset or before dawn. It's much closer to the Sun than we are, so it never appears to be too far east or west of the Sun.

MERCURY WITH BINOCULARS

Mercury is too small to see clearly with binoculars, but they will help to make it appear much brighter against the background sky.

MERCURY WITH A TELESCOPE

With a telescope, you can zoom in and see the phases of Mercury clearly. When it is on the far side of the Sun, it will appear as a gibbous phase (greater than a half-circle). When it is on the near side of the Sun, it can be seen as a crescent shape.

Filters won't make your view of Mercury much clearer.

VICIOUS VENUS

Venus is often thought of as Earth's evil twin! It is one of the most undesirable places anywhere in the Solar System. Covered with volcanos, Venus is blanketed with a thick atmosphere. Its surface is astonishingly hot, day or night, and it experiences crushing pressure, with droplets of acid in the air! Humans wouldn't last long on Venus. Fortunately, you can admire it from afar.

 DID YOU KNOW?

If all the planets were lined up at their average distances from the Sun, Venus would be the nearest to the Earth. Venus also spins clockwise, which is the opposite way to all the other planets in the Solar System!

FACTFILE

Diameter: 12,104 km (slightly smaller than Earth)
Orbital period: 225 days (0.6 years)
Length of a solar day: 117 Earth days
Average distance from the Sun: 108.2 million kilometres
Number of moons: 0
Strength of gravity: 90% of Earth's gravity

THE MORNING STAR

Venus is a dazzling bright planet when viewed with just your eyes. After the Sun, Moon and International Space Station, it is the fourth brightest object in the sky. It is often called 'Morning Star' or 'Evening Star' because it is regularly visible in the morning or evening.

VENUS WITH BINOCULARS

You can see the phases of Venus with binoculars.

VENUS WITH A TELESCOPE

A telescope will show you the phases of Venus very clearly, but it's virtually impossible to see any features. Venus does have some clouds, but they are too pale to be seen without using a filter.

Amber (#15), red (#25) or violet (#47) filters may help you to identify clouds on Venus, but these are rarely visible and usually require a large telescope with a high magnification to see clearly. If you have any of these filters in your collection, it's always worth having a go.

MYSTERIOUS MARS

The planet Mars is also known as the 'Red Planet'. While its reddish-orange colour makes it appear warm, it is actually a freezing cold desert world. Mars is smaller than Earth, and appears small to us in the sky, but it has a variety of surprises.

FACTFILE

Diameter: 6,792 km (about half the size of Earth)
Orbital period: 687 days (1.9 years)
Length of day: 24.7 hours
Average distance from the Sun: 228 million kilometres
Number of moons: 2
Strength of gravity: 38% of Earth's gravity

IS THERE LIFE ON MARS?

There's probably not life on Mars today. Mars, however, was once quite similar to Earth – a water world with a warmer climate – so perhaps something *did* live on Mars a long time ago. It's certainly mysterious enough to spark the curiosity of scientists, who have sent many robots there over the years to investigate its past.

MARS WITH BINOCULARS

Mars appears very small to us on Earth, but with a pair of binoculars you will be able to tell that it is not a star. You will see its striking red-orange colour clearly. This colour comes from rusting iron in the planet's soil.

MARS WITH A TELESCOPE

You need a high magnification to see Mars clearly. The Martian landscape is full of surprises. You might be able to pick out one of the planet's icy polar caps, appearing as a bright spot on the edge of the disk. The large dark marks of Syrtis Major (a region of cooled volcanic lava) can be seen; you may even spot clouds forming over the giant Mount Olympus and Tharsis Mountains.

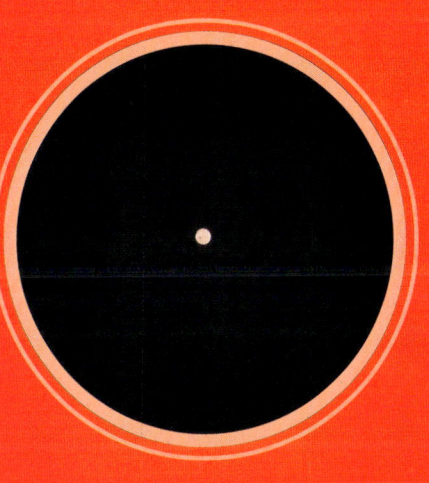

A yellow (#12) filter can make surface features easier to pick out, while an amber (#15) filter will make the ice caps stand out more clearly. A light red (#23A) filter works very well for all Martian features. With larger telescopes, a light blue (#82A) filter can make dark features more obvious, and a light green (#56) filter brings out the polar ice caps.

JUMBO JUPITER

Jupiter is the largest planet in the Solar System. This planet is a giant ball of gas, with no solid surface. Jupiter has many moons orbiting it. The Galilean Moons are four particularly spectacular moons called Io, Europa, Ganymede and Callisto. They were discovered by Galileo Galilei. Io is a volcanic moon and the other three are icy, with oceans beneath their surfaces.

FACTFILE

Diameter: 142,984 km (more than 11 times the size of Earth)
Orbital period: 4,333 days (11.9 years)
Length of a solar day: 9.9 hours
Average distance from the Sun: 778.5 million kilometres
Number of moons: 95 (currently)
Strength of gravity: 240% of Earth's gravity

JUPITER WITH BINOCULARS

Jupiter shows as a small disk in binoculars. All four of the Galilean Moons can be seen alongside it. With just your binoculars, you can see moons orbiting another planet!

JUPITER WITH A TELESCOPE

Jupiter really comes to life with a telescope. It has no solid surface but layers of fast-moving clouds that are full of giant storms. The most famous of these is the Great Red Spot, which is larger than planet Earth. In a telescope, it appears pale orange in colour.

The dark orange stripes on Jupiter are called belts, while the lighter bands are called zones. With a larger telescope, you can see knots and strange oval-shaped storms called barges inside the belts. You may also see blue festoons (smaller storms) on the edges of the belts.

Orange (#21), dark blue (#38A) and blue (#80A) filters all make the belts and Great Red Spot easier to see.

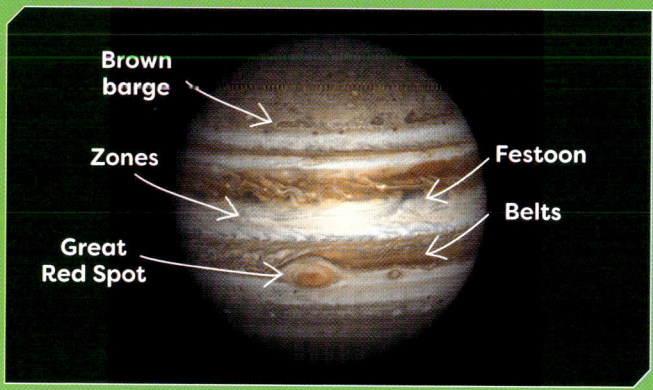

SERENE SATURN

Saturn appears to be less dramatic than the other planets, which is why it is described as serene. For most astronomers, Saturn is the jewel of the Solar System thanks to its remarkable bright rings. These rings are made up of tiny pieces of water ice. They are extremely shiny and reflect sunlight well, making them very easy to observe from Earth.

WHAT IS SPECIAL ABOUT SATURN'S MOONS?

Saturn has the most moons of any other planet in the Solar System. Its largest moon, Titan, is unique among Saturn's moons because it is the only moon in the Solar System with an atmosphere, and it looks more like a planet. Titan is very cold, but it has lakes of liquid methane on its surface.

Diameter: 120,536 km (more than 9 times the size of Earth)
Orbital period: 10,747 days (29.4 years)
Length of a solar day: 10.6 hours
Average distance from the Sun: 1.4 billion kilometres
Number of moons: 146 (currently)
Strength of gravity: 107% of Earth's gravity

FACTFILE

SATURN WITH BINOCULARS

Binoculars will show that there is something unusual about Saturn's shape but you'll need a telescope to make out its rings.

SATURN WITH A TELESCOPE

Saturn is a treat to see in a telescope: a pale ball suspended in a perfect ring. There are actually 7 main rings around Saturn, named from A to G. The A ring and the B ring are the largest, and they are separated by a gap called the Cassini Division. With high magnifications, it is possible to see this gap as a darker region within the rings.

A yellow-green (#11) filter will make the Cassini Division appear darker.

ISOLATED ICE GIANTS

Saturn is the last planet which is easily visible to the eye, but beyond it lie two more remote worlds. They were not discovered until after the telescope was invented, even though one of them is just visible to the eye on a very dark night. They are the ice giants Uranus and Neptune.

URANUS

FACTFILE

Diameter: 51,118 km (roughly 4 times the size of Earth)

Orbital period: 30,589 days (83.8 years)

Length of a solar day: 17.2 hours

Average distance from the Sun: 2.9 billion kilometres

Number of moons: 28 (currently)

Strength of gravity: 90% of Earth's gravity

NEPTUNE

FACTFILE

Diameter: 49,528 km (almost 4 times the size of Earth)

Orbital period: 59,800 days (163.7 years)

Length of a solar day: 16.1 hours

Average distance from the Sun: 4.5 billion kilometres

Number of moons: 16 (currently)

Strength of gravity: 112% of Earth's gravity

FAR FAR AWAY

Uranus and Neptune are extremely far away from the Sun. They receive very little light, and they are very cold. From Earth, both planets appear to be very small and faint. They are similar in colour, with Uranus appearing pale green and Neptune more blue. Uranus and Neptune are very challenging to observe in detail.

URANUS AND NEPTUNE WITH BINOCULARS

Both Uranus and Neptune appear like stars in binoculars, and their disks are very small when seen from Earth.

URANUS AND NEPTUNE WITH A TELESCOPE

A telescope with a high magnification will give you a great view of Uranus. It looks like a little pea in front of the stars. Uranus has many moons, a few of which are visible with large telescopes.

Neptune looks slightly smaller in a telescope than Uranus because it is farther away. In fact, Neptune and Uranus are almost the same size. Neptune's largest moon, Triton, can be seen with a large telescope. It is easy to mistake Triton for a star because it often seems to be so far away from Neptune.

ASTEROIDS

The asteroid belt is a region of space between the orbits of Mars and Jupiter that is home to a large number of asteroids. Asteroids are tiny fragments of rock and metal left over from when the Solar System was formed. They are sometimes called minor planets.

DID YOU KNOW?

The word 'asteroid' means 'star-like'.

Asteroid belt

Sun

DANGEROUS VISITORS

Asteroids can sometimes drift towards Earth and even crash into it. Fortunately, most are too small to reach the ground without burning up! Those that do reach the ground are often too small to do any damage, and large impacts are rare. The last one happened 66 million years ago, and led to the extinction of most of the dinosaurs!

ASTEROIDS TO LOOK FOR

Asteroids appear to us like stars, even with a telescope. These are some of the brightest asteroids in the Solar System. The largest of all is Ceres, which is now considered to be a dwarf planet.

Name	Diameter
Ceres	952 kilometres
Vesta	530 kilometres
Pallas	545 kilometres
Iris	200 kilometres
Hebe	185 kilometres
Juno	234 kilometres
Melpomene	140 kilometres
Eunomia	255 kilometres
Flora	136 kilometres
Metis	190 kilometres

Asteroid Vesta

COMETS

Comets are icy visitors in the sky. Many of them come from a place beyond the planets far away in the Solar System. Comets are minor planets, but they are not the same as asteroids. While asteroids are made of rock and metal, comets are mostly ice – frozen gases and water.

Comets travel around the Sun following an oval-shaped path. When a comet finds its way into the inner Solar System, it is heated up by the Sun's energy.

The comet begins to break apart, leaving a large tail of icy particles behind it. Comet tails can be astonishingly long – hundreds of millions of kilometres. Some 'great comets' produce tails that stretch across huge sections of the sky!

 DID YOU KNOW?

The word comet means 'hairy' or 'bearded' star. Our ancestors had no way of knowing what the tails of comets were really made of!

WHY ARE COMETS GREEN?

Comets are one of the few things in the night sky that can appear green. Their atmospheres contain carbon, which absorbs ultraviolet (UV) light from the Sun and releases it as a green glow.

The arrival of a great comet is impossible to predict. It can happen without much warning, as comets are hard to detect when they are too far from the Sun.

Comets are given a number and named after the person or spacecraft that discovered them. In 2020, a comet called NEOWISE produced a beautiful tail that was visible even in the brightly lit skies of London!

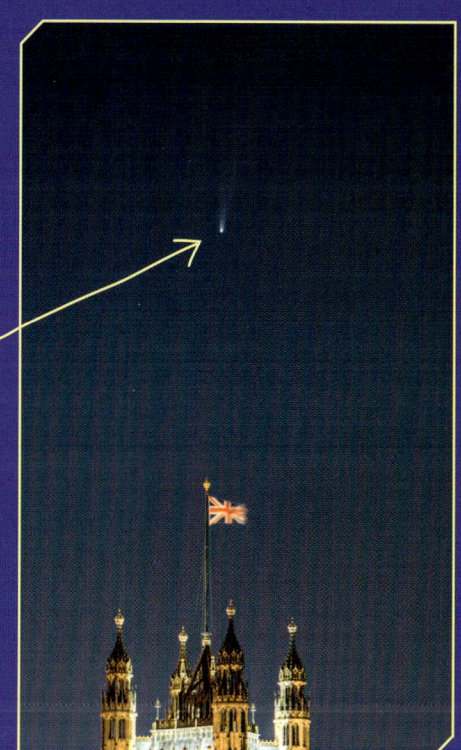

SHORT-PERIOD COMETS

While long-period comets can take up to 30 million years to complete their orbit, short-period comets take less than 200 years to orbit the Sun. Some take much less than this and visit the sky above Earth every few years. Here are a few of the best ones to look out for:

Comet number and name	Length of orbit
8P/Tuttle	13.5 years
46P/Wirtanen	5.5 years
6P/d'Arrest	6.6 years
41P/Tuttle-Giacobini-Kresák	5.4 years
2P/Encke	3.3 years
36P/Whipple	8.5 years

METEOR SHOWERS

Meteors (sometimes called shooting stars) are bright streaks of light that are seen when meteroids burn up in the Earth's atmosphere. Meteoroids are tiny pieces of rock.

When we see a number of meteors at once, it is known as a meteor shower. Meteor showers often occur when the Earth passes through the debris from a comet's tail.

 DID YOU KNOW?

Meteor showers appear to come from a particular constellation – this is how they get their names. For example, the Orionid meteor shower appears to come from the constellation Orion. Meteor showers are shown on the star charts on pages 46–83 with this symbol.

METEOR SHOWER DATES

Name	Dates
Quadrantids	1–10 January
Lyrids	16–25 April
Eta Aquariids	19–26 May
Alpha Capricornids	11 July–10 August
Delta Aquariids	12 July–23 August
Perseids	13 July–26 August
Alpha Aurigids	August–October
Southern Taurids	7 September–19 November
Orionids	4 October–14 November
Northern Taurids	19 October–10 December
Leonids	5–30 November
Geminids	4–16 December
Ursids	17–23 December

The purple rows in the table show the strongest meteor showers.

FEROCIOUS FIREBALLS

The brightest meteors are known as fireballs. Fireballs are more common during meteor showers. They can show bright colours and leave trails in the sky that take seconds or even minutes to fade away. The colour we see from a bright meteor depends on what the rock is made from.

Colour	Made from
Cyan	Magnesium
Red	Nitrogen/oxygen
Yellow	Iron
Violet	Calcium
Amber	Sodium

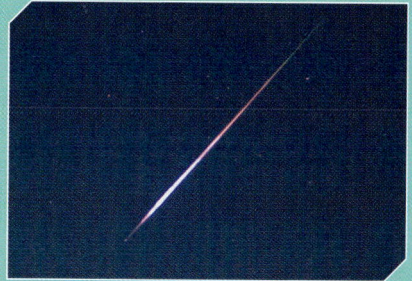

THE INTERNATIONAL SPACE STATION (ISS)

Would you like to wave to an astronaut? Every 90 to 93 minutes, a crew of astronauts orbits the Earth on board the International Space Station (ISS). Go outside and see it in the sky for yourself!

The ISS is a satellite that is in orbit around the Earth. It is a place where astronauts are learning to live and work in space. It's full of science experiments and instruments.

To generate its power, the ISS has two very large solar panels. In certain positions, they reflect sunlight down towards the ground, making the Station appear as a bright star sailing across the sky.

FACTFILE

Year launched: 1998
Length of orbit: 90–93 minutes
Mass: 420,000 kilograms
Width: 109 metres
Altitude: 400 kilometres
Typical crew: 7

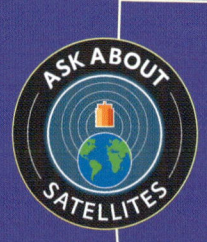

HOW CAN I SEE THE ISS?

Ask a grown-up to help you use NASA's 'Spot the Station' service to predict when the ISS will be visible to you (see the **Resources** section). The ISS's journey begins in the west and moves towards the east. The Station looks like a very bright star, which eventually fades away as it moves into the shadow of the Earth.

MORE SATELLITES

The International Space Station is not the only satellite orbiting the Earth. In fact, there are thousands of satellites in space. They are used in lots of ways such as communication, navigation and weather forecasting.

Using the resources found at the back of this book, you can look up which satellites are visible and find out which ones you've seen. Make sure you note down the exact time that you see a satellite and which constellation it is visible in. You will need this information to find out the name of the satellite.

SATELLITES VERSUS THE NIGHT SKY

While it can be exciting to see satellites travelling overhead, they contribute to the light pollution of the sky which can make astronomy more challenging. Some astronomers think the beauty of the sky is at risk. Other people believe that the uses of satellites are more important than stargazing. What do you think?

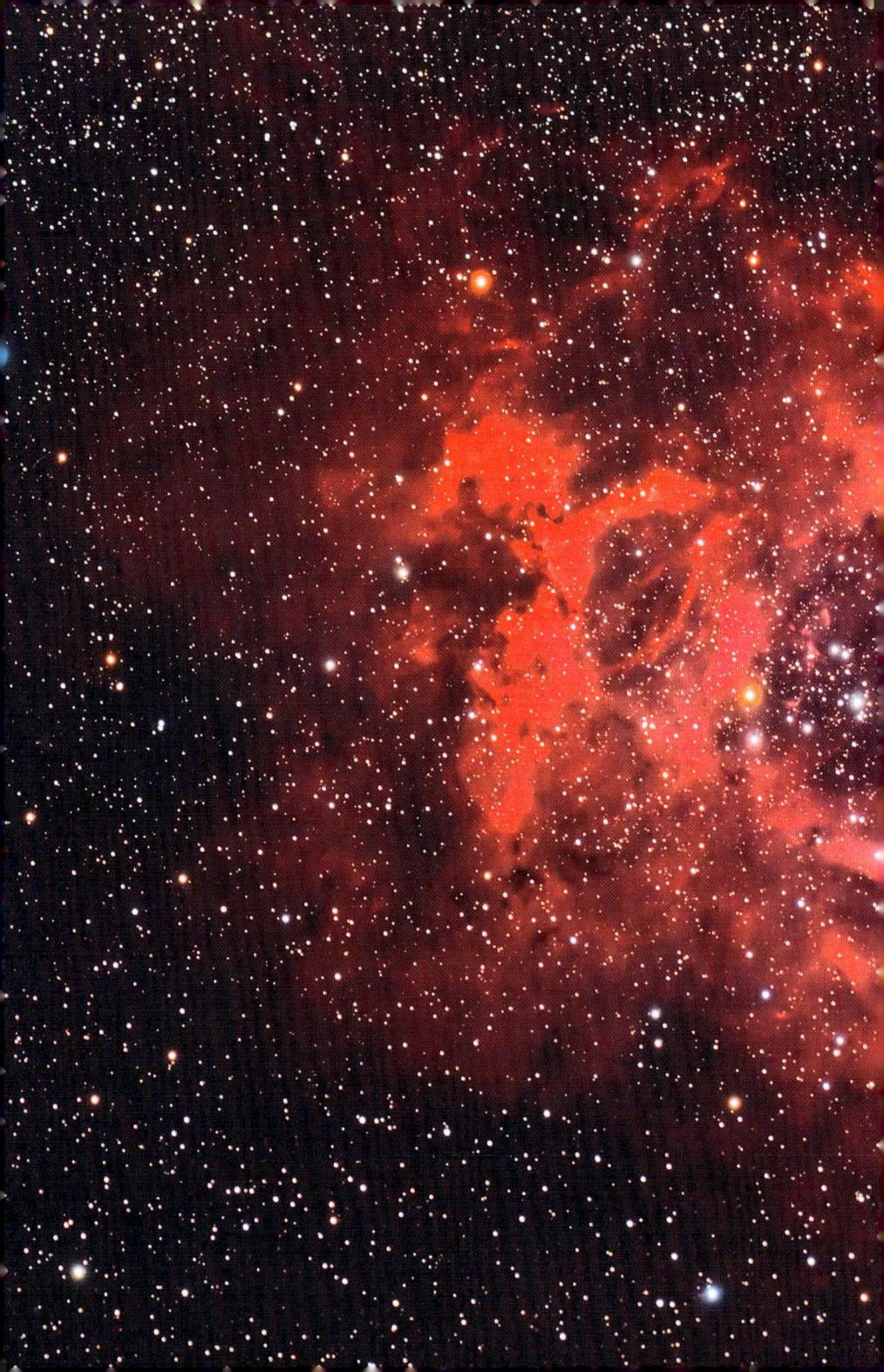

THE MILKY WAY

Beyond the Solar System, almost everything else we see in the night sky is a part of the Milky Way, and so are you! The Milky Way is our home galaxy, just one of many billions of galaxies in the Universe.

The Milky Way is home to stars, star clusters and nebulae. These belong to a category called deep sky objects. They are often challenging but very rewarding to see with your own eyes.

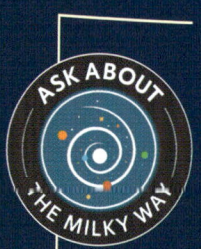

WHERE ARE WE IN THE MILKY WAY?

We live about halfway out from the centre of the Milky Way, among the spiral arms of stars and gas.

You are here!

❓ DID YOU KNOW?

The Milky Way's name refers to its milky appearance in the sky. In ancient Greek, the Milky Way was called 'galaxias', (meaning 'milky') and this is where the word 'galaxy' comes from.

WHAT TO LOOK FOR IN THE MILKY WAY

The Milky Way appears as a band of faint light in the sky. This is the light of many billions of stars. There are so many stars, and they are so far away that they appear together like a glowing cloud.

The Milky Way appears brightest in the summer, when we can look both towards its centre and along the part of the spiral arm that we live in. In the winter, when the Earth has moved halfway around the Sun, we can still see the Milky Way, but we are looking out of the Galaxy, away from the centre. This is why it appears fainter, even though the nights are darker.

THE MILKY WAY CENTRE

To find the centre of the Milky Way, look towards the Teapot asterism in the constellation Sagittarius on summer nights (see pages 25 and 62–63).

THE CYGNUS STAR CLOUDS

The Cygnus star cloud is a bright patch of the Milky Way, where there are millions of stars. It is found in the constellation Cygnus, in the northeast of the Summer Triangle, surrounding the star Deneb.

THE MILKY WAY WITH BINOCULARS AND TELESCOPES

The Milky Way is rich in stars, clusters and nebulae, many of which can be seen clearly through binoculars or a telescope. When you look towards the Milky Way, you will discover many glittering regions and glowing patches. Take your time to explore it on a dark night and enjoy the view.

WHY ARE THERE 'HOLES' IN THE MILKY WAY?

There are dark patches in the Milky Way where it seems there are very few or no stars. Although they look like holes, these dark patches are actually large clouds of gas within the galaxy. These clouds block out the light of the stars behind them.

OPEN STAR CLUSTERS

Stars are not spread out evenly in the Milky Way. Many of them are grouped together in clusters. Open clusters, also known as galactic clusters, are common within the Galaxy's spiral arms.

Open clusters can sometimes be thought of as families, because the stars were born together from the same cloud of gas. They are siblings travelling through the Milky Way, and the force of gravity keeps them together.

A few star clusters are visible with just your eyes but lots more can be seen through binoculars and telescopes. How many of these can you find?

PLEIADES CLUSTER (M45)

This is the most famous star cluster in the sky and is found in the constellation Taurus. It is also known as the Seven Sisters. It looks best in binoculars or a telescope with very low magnification.

HYADES CLUSTER (C41)

After the Pleiades, the Hyades is the other popular star cluster in the constellation Taurus. It is shaped like a letter V and makes up the head of the bull.

Open star clusters are shown on the star charts with this symbol on pages 46–83

DOUBLE CLUSTER (C14)

Two clusters share the sky between the constellations Cassiopeia and Perseus. They are both just visible to the eye, even at a distance of about 7,500 light years!

WILD DUCK CLUSTER (M11)

Can you see a flock of flying ducks when looking at this cluster in the constellation Scutum?

SPIRAL CLUSTER (M34)

Stars seem to spiral out from the centre of this sparse but pretty cluster in the constellation Perseus.

BEEHIVE CLUSTER (M44)

The Beehive Cluster sparkles in the heart of the faint zodiacal constellation Cancer.

127

GLOBULAR CLUSTERS

Globular clusters are enormous, sphere-shaped collections of stars. Some globular clusters contain millions of stars.

Unlike open clusters, which are found in the Milky Way's spiral arms, globular clusters are found in the region around the Galaxy, known as the Milky Way's halo. Globular clusters are generally further away than other deep sky objects in the Milky Way. Each globular cluster has a brighter core surrounded by a fainter halo of stars that are spread more widely apart.

HERCULES CLUSTER (M13)

Look to the Keystone asterism in the constellation Hercules to find one of the best-known globular clusters. The Hercules Cluster is home to about half a million stars and is faintly visible to the eye. You can see it high up in the sky on summer nights.

 Globular clusters are shown on the star charts with this symbol on pages 46–83

PEGASUS CLUSTER (M15)

Over 100,000 stars make up the Pegasus Cluster, in the constellation Pegasus. It hangs in front of the flying horse's nose like a dangling carrot!

M4

M4 is the closest globular cluster to the Sun – it is about 6,000 light years away. It looks big in the sky and is a very nice sight through binoculars. M4 appears low in the summer night sky, very close to the star Antares (the heart of the constellation Scorpius).

M22

M22 is just above the Teapot asterism. It can be quite challenging to find because it doesn't rise very high above the horizon from most places in the northern hemisphere. However, M22 is large and bright, and still very impressive with binoculars or a telescope.

MULTIPLE STAR SYSTEMS

When you look at stars through a telescope, you will sometimes find that they have a companion star that you can't see with the unaided eye. Double stars are made up of two stars and multiple star systems are made up of three or more stars.

When a telescope reveals enough detail to separate double or multiple stars, it is known as 'splitting'. Can you split these stars?

MIZAR

Mizar is a member of the Plough (Big Dipper) asterism. Look carefully and, with just your eyes alone, you'll see it has a companion called Alcor. This is one of the widest double stars in the sky. The names Mizar and Alcor mean 'horse' and 'rider'.

ALBIREO

Albireo is a beautiful double star. Its two members are golden and blue in colour, and very noticeable in a telescope. Try adjusting the telescope so that it is slightly out of focus. This will spread out the colourful light and make it easier to see. Albireo is in the beak of the swan in the constellation Cygnus.

CASTOR

Castor is one of the twins of Gemini. A telescope can show you three of the six stars in this system. It is well worth a look.

ALMACH

For some, Almach in the constellation Andromeda is even more impressive than Albireo. It appears as a yellow-orange star and a blue star, close together. The blue star has two more companions we don't see, so there are four stars in total. Which do you prefer, Almach or Albireo?

EPSILON LYRAE

In the constellation Lyra, close to the star Vega, lies this 'double double' star. Epsilon Lyrae is easy to split into two stars called ε^1 and ε^2 with a telescope, but look closer: each star is in fact a pair on its own! You'll need a high magnification and steady skies to make them out.

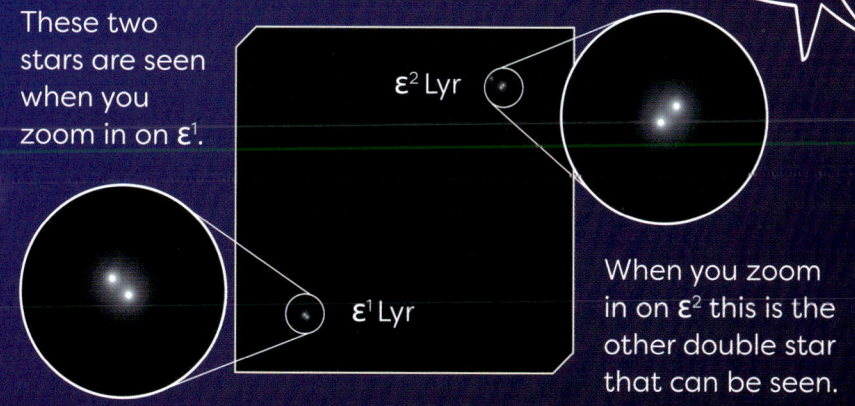

These two stars are seen when you zoom in on ε^1.

ε^2 Lyr

ε^1 Lyr

When you zoom in on ε^2 this is the other double star that can be seen.

NEBULAE

The word 'nebula' means 'cloud'. Clouds are found across the Galaxy, but they're not like the kind that form in our skies on Earth. Nebulae are large clouds of gas that are usually many times larger than our entire Solar System. There are a few kinds of nebulae that can be seen in the night sky.

EMISSION NEBULAE

Energy released from the stars within these nebulae makes them glow. Many of the brightest emission nebulae we see are star nurseries, where young stars begin their lives. The next page shows you some of the most impressive emission nebulae to look for in the night sky.

REFLECTION NEBULAE

Sometimes the light of nearby stars reflects off a nebula and makes it glow. This is known as a reflection nebula. The bright Pleiades Cluster (see page 126) does this, giving the nebula a ghostly appearance in photographs.

DARK NEBULAE

When a nebula is seen in silhouette, blocking out the light from behind it, it takes on the appearance of a hole in the sky. This is called a dark nebula.

 Emission nebulae are shown on the star charts with this symbol on pages 46–83

ORION NEBULA (M42)

This is a large star nursery where solar systems are being born. You can see it with just your eyes in the middle of the sword in the constellation Orion. With a telescope, you can find a group of stars at its centre. The four brightest stars form a pattern called the Trapezium.

LAGOON NEBULA (M8)

This nebula in the constellation Sagittarius lives up to its name! It looks calm, like a tropical pool.

ROSETTE NEBULA (C49)

This cloud of gas looks like a flower, and it surrounds a large star cluster. It appears quite faint so it's challenging to see. You'll find it in the constellation Monoceros, clearest on a winter night.

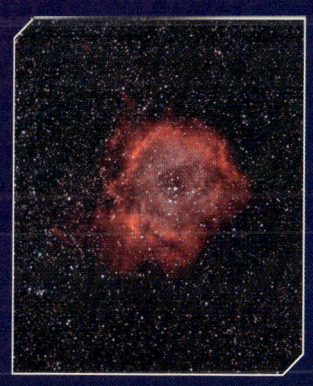

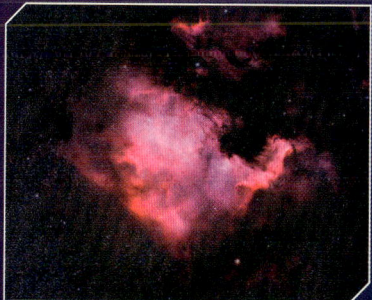

NORTH AMERICA NEBULA (20)

In the constellation Cygnus, you'll find this large nebula, which is shaped a bit like the continent of North America.

PLANETARY NEBULAE

When the Sun reaches the end of its normal life, it will transform into a planetary nebula. We can't know exactly what it will look like because planetary nebulae are unique! Many other stars which were once like the Sun have formed nebulae across the Milky Way.

 DID YOU KNOW?

Planetary nebulae are sometimes called 'planetaries' for short, but they aren't related to planets at all. The name was introduced in the eighteenth century by astronomers who compared their round shapes to planets.

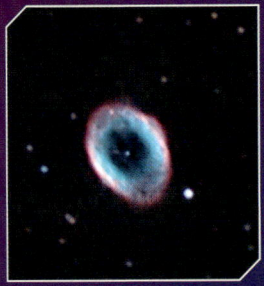

RING NEBULA (M57)

The Ring Nebula in the constellation Lyra looks like a little ring of smoke in a telescope. Only a large telescope will show you the star in the centre, which produced the nebula before shrinking to become a tiny white dwarf star.

HELIX NEBULA (NGC 7293)

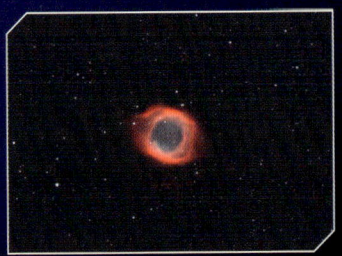

The Helix Nebula can be found in the constellation Aquarius. It looks similar to the Ring Nebula but is fainter and its edges are fuzzier.

 Planetary nebulae are shown on the star charts with this symbol on pages 46–83

OWL NEBULA (M97)

Through a large telescope, the Owl Nebula looks like a round face with two large eyes, just like an owl. With binoculars or a small telescope, these 'eyes' are difficult to spot, but the nebula itself is visible as a little round patch of light. It is in the constellation Ursa Major.

DUMBBELL NEBULA (M27)

The dumbbell shape of this nebula in the constellation Vulpecula can be faintly seen with a telescope. It looks a little like a bow tie.

NGC 2392

You will need a larger telescope to see this nebula. Hidden away in the constellation Gemini is a cute little nebula with a brighter inner region and fainter outer region.

BLINKING PLANETARY NEBULA (C15)

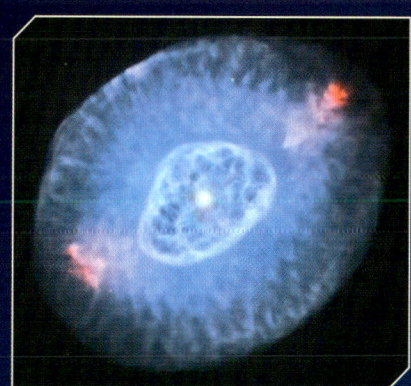

The Blinking Planetary in the constellation Cygnus is famously strange. When you look directly at it, the nebula seems to disappear. Only the star at the centre is visible, but it can be viewed well using averted vision (see page 41). When you look to one side, the nebula comes back into view! Its name comes from this odd behaviour.

OTHER GALAXIES

There are several types of galaxies. Some are spiral galaxies (like the Milky Way) and others are known as elliptical galaxies. The galaxies described here are all visible with just binoculars, despite being millions of light years away.

ANDROMEDA GALAXY (M31)

Our nearest neighbour galaxy is so bright it can be seen with just your eyes. At 2.5 million light years away, it is the furthest thing you can see with your eyes. We're only able to see the bright centre, but if the entire galaxy was visible, it would appear to be six times wider than the Full Moon! Found in the constellation Andromeda, it has two notable satellite galaxies that are visible in binoculars and telescopes.

WHIRLPOOL GALAXY (M51)

This magnificent spiral galaxy is popular with stargazers using telescopes, but it can also be seen in binoculars. It interacts with another small galaxy, and so it appears as two smudges of light rather than one. It is very close to the Plough (Big Dipper) asterism.

 Galaxies are shown on the star charts with this symbol on pages 46-83

BODE'S GALAXY AND THE CIGAR GALAXY (M81 AND M82)

In the constellation Ursa Major, two galaxies appear close together in the sky. Bode's Galaxy is a spiral, seen almost face-on, while its neighbour the Cigar Galaxy is seen from the side. Bode's Galaxy looks quite round and the Cigar Galaxy quite narrow. Together they make a lovely pair in binoculars and telescopes with low magnification.

SOMBRERO GALAXY (M104)

This is an edge-on spiral galaxy in the constellation Virgo that astronomers love to find in the spring. In binoculars, it is a small, round patch of light. Telescopes can reveal a longer, thin oval with a dark line cutting through it. This is the shadow of the gas inside the spiral arms of the galaxy.

M87

While this galaxy only appears as a faint smudge with no features, it is a remarkable object. It has no spiral arms but it contains a large number of stars – probably several trillion! You can find this one in Virgo on dark spring nights.

GALAXY CLUSTERS

Galaxies, like stars, can form into clusters. Our own Milky Way is part of the Virgo Supercluster, which contains many smaller groups of galaxies. Some astronomers call spring 'galaxy season' because on spring nights, we can look towards the constellations of Leo and Virgo and find a large number of galaxies to explore.

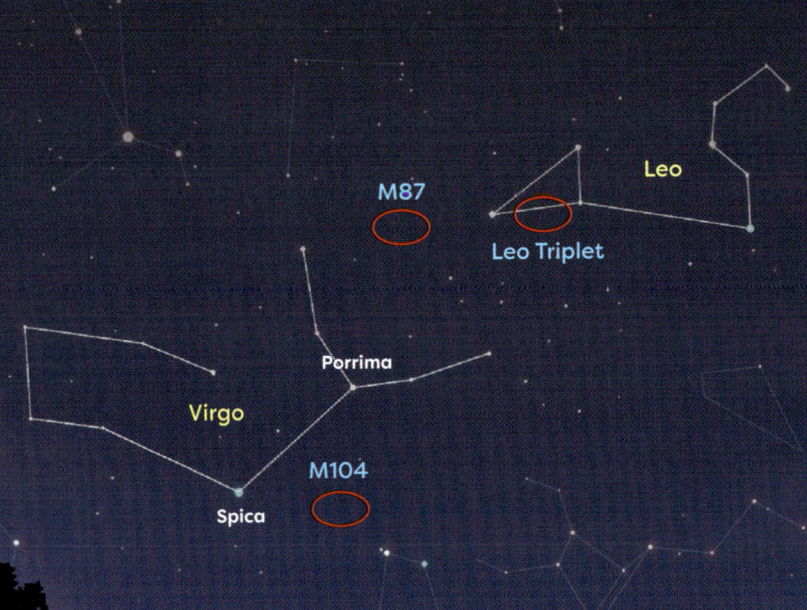

LEO TRIPLET

The Leo Triplet is made up of three galaxies called M65, M66 and NGC 3628. All three are spiral galaxies, and all three are visible in telescopes and binoculars. To find them, look just beneath the back legs of the lion in the constellation Leo.

VIRGO CLUSTER

The Leo Triplet, as well as many other groups of galaxies, is part of the Virgo Cluster. Looking to the east of Leo, towards Virgo, you can gaze into a part of the sky that has many galaxies. Scan the area between the constellations Virgo and Leo to find the distant M87 (see page 137), and use binoculars or a telescope to scan between the bright stars Spica and Porrima to look for M104 – the Sombrero Galaxy (pages 54–55 and 134).

AMAZING AURORAS

The Earth creates beautiful displays of light in rings around its north and south poles. These natural lights dance in the sky and flicker with colour. They are called auroras.

Auroras in the Northern Hemisphere are called the Northern Lights, also known as 'aurora borealis'. They form when particles flowing out from the Sun become trapped inside the Earth's magnetic field. These particles have a huge amount of energy and, as they plunge down into the atmosphere, they give that energy to gas atoms and molecules in our atmosphere. This causes them to glow and give off different-coloured lights.

WHERE AND WHEN CAN I SEE THE NORTHERN LIGHTS?

On rare occasions, the Northern Lights can be seen outside the Arctic in parts of northern Europe, the UK and northern United States.

Alaska, parts of Canada, Greenland, Iceland, Norway, Finland, and Sweden are great places to see the Northern Lights. They are visible in the night sky throughout the northern winter, between the months of September and March. They're best seen around the new moon when there are darker night skies.

❓ DID YOU KNOW?

The word 'aurora' means 'dawn'. Galileo Galilei came up with the term 'aurora borealis', meaning 'northern dawn', in 1619.

STRANGE CLOUDS AND ICY SIGHTS

You usually need a sky free of clouds for stargazing, but there are some interesting rare clouds to look out for in the sky.

NOCTILUCENT CLOUDS

Noctilucent clouds can occasionally be seen on summer evenings. They appear just after sunset when the brightest stars are beginning to become visible. These clouds are so high up, in the colder part of the atmosphere, that they are made up of ice crystals. Noctilucent clouds are silvery in colour, and they may also shine pale blue.

LUNAR HALO

Also caused by ice crystals high in the atmosphere, the Moon can also produce a halo on very cold nights. This happens when the Moon is at a bright phase, such as gibbous moon or full moon. Lunar halos are very striking!

SUN DOGS

Have you ever seen a halo of light around the Sun? When the air is very cold, ice crystals produce this halo. On the left and right side of the Sun, you can sometimes see two bright spots in the halo known as sun dogs. These occasionally shine with all the colours of the rainbow.

NACREOUS CLOUDS

Nacreous clouds are another type of icy cloud. They are most commonly seen in winter time and usually in the far north, such as the Arctic. The sunlight spreads out as it passes through them and creates beautiful colours.

COLOURS OF THE ATMOSPHERE AND COSMIC DUST

The white light from the Sun is made up of lots of different colours. Sunlight is scattered (directed in different directions) when it passes through the Earth's atmosphere. Blue light is scattered in every direction, making the sky appear blue.

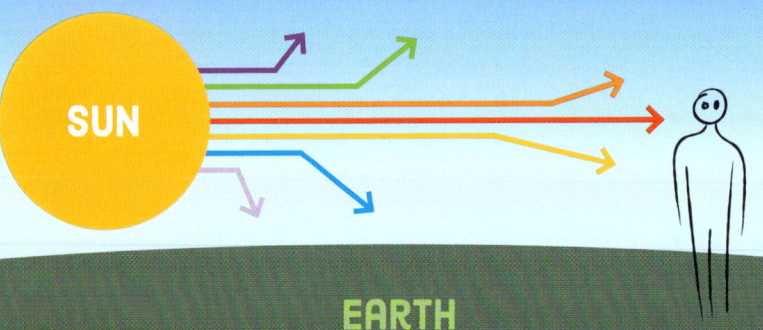

THE BELT OF VENUS

When the Sun is low over the horizon, around sunrise or sunset, the sunlight has further to travel through the atmosphere. The blue light is scattered away and we see the yellow and red light. This makes the sky look orange. If you look away from the setting or rising Sun, you'll see the red light reflecting off the sky as a pink band. This is called the Belt of Venus.

ZODIACAL LIGHT

If you visit a dark stargazing site with perfectly clear skies, look west after sunset to see a faintly glowing triangle of light over the horizon. It's called zodiacal light. What you can see is sunlight reflected by dust grains throughout the Solar System. This is very rare to see!

ASTRO PHOTOGRAPHY

ASTROPHOTOGRAPHY WITH CAMERAS AND SMARTPHONES

Would you like to take your own amazing photos of the night sky? Astrophotography is easier than ever! Today's cameras and smartphones are very good at capturing starlight, showing the colours and details of stars, nebulae and the Milky Way above us.

USING A SMARTPHONE

Modern smartphones have become great astrophotography cameras. Simply point a smartphone at the sky, wait for a moment, and it will automatically adjust itself for the conditions.

In some cases, you may need to enable a night mode in the phone's camera app. The icon often looks like this.

USING A CAMERA

The night sky is dark, and taking a good picture of it requires you to use a long exposure. This means the camera's shutter is open for a long time. While the shutter is open, the camera will collect light. Remember, the more light you can collect, the brighter an object will appear. Many cameras have a night shot mode which sets a long exposure.

SHUTTER BUTTON

FOCUS RING

LENS

ASTROPHOTOGRAPHY ACCESSORIES

There are some handy gadgets that can help with your photos of the night sky.

INTERVALOMETER

An intervalometer is a device that plugs into the camera and automatically takes sequences of photos. This is helpful for capturing photos of meteor showers where you have to be patient to spot them. With an intervalometer you can set up your camera, telling it how many images to take and how often – then set it going and sit back! Your camera may already include an intervalometer mode though, so check before buying something you don't need!

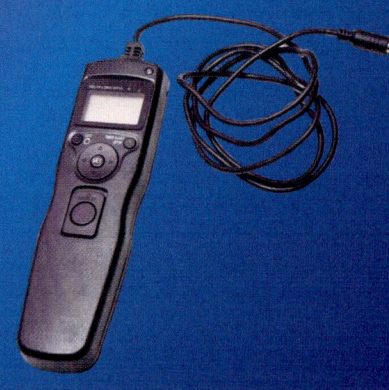

LENSES

Astrophotographers use lenses that are 'fast'. They collect a lot of light and this lets you take bright pictures in a short time. This means you will get a clearer photograph.

TRIPOD

A tripod is very important for astrophotography. It is a three-legged stand that keeps the camera steady while the shot is being taken. It's impossible to hold a camera steady with your hand for more than a fraction of a second, and to photograph the night sky, you'll need to take shots that last several seconds.

If you're using a smartphone, you will need a smartphone tripod adapter to attach your phone to it.

ASTROPHOTOGRAPHY TIPS AND TRICKS

Follow these tips and tricks and become an expert astrophotographer in no time!

PRACTISE SETTING UP EVERYTHING INDOORS

It can be tricky to set things up when you're out in the dark, so practise indoors first.

MAKE SURE EVERYTHING IS STURDY

Attach the camera or smartphone to the tripod carefully and ensure it's very sturdy. When the air gets cold, screws can become loose, so check everything is tight.

SHOOT IN THE HIGHEST QUALITY YOU CAN

You may be able to take high-quality images, called raw files. Raw files are much better for editing later on.

USE A TIMER

Use the camera's two-second countdown timer to avoid the camera's shake making your photos blurry.

SHOOT, ADJUST, SHOOT AGAIN

Astrophotography involves trial and improvement. Take a shot and look at it on the screen. What can you change to improve it? Try altering the settings on the camera and take another shot.

EDIT LATER

You can adjust your images using image processing software, but don't edit them too much. If you try to change the colour, contrast or detail too much, it will look unnatural.

ASTROPHOTOGRAPHY **TARGETS**

If you're wondering where to start when it comes to photographing the night sky, why not have a go at making your way through this list of targets?

CONSTELLATIONS

Use a wide lens to capture constellations like Orion. Their bright stars will make them easy to find in your image. A camera can show us much more than the eye would see.

A long lens or a telescope, that make distant objects appear magnified, can be used to capture details in the Orion Nebula.

MILKY WAY

In the summer, point a camera towards the Summer Triangle or the Teapot Asterism and take a long exposure photograph to capture the faint light of billions of distant stars in the Milky Way. You can even capture our galaxy with a smartphone using night mode.

THE MOON

The Moon is very bright and you can take short-exposure photographs to reveal details on its surface. Try attaching a camera or smartphone to a telescope to take a close-up picture like this one. You'll be able to see many craters, mountains and valleys. You might even capture some faint colour on the Moon's surface!

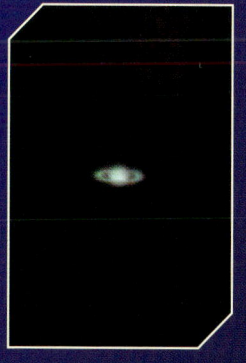

PLANETS

Planets always appear small, even in telescopes. In most pictures, they look like any other star. If you carefully hold a camera or smartphone in front of the eyepiece of a telescope, you can capture photos that show details on the planets. Here you can see Saturn's rings as they appeared through the telescope.

STAR TRAILS

With very long exposures, or by joining together lots of shorter exposures, you can capture 'star trails' which reveal the rotation of the Earth. Point the camera towards the true north (see page 28) to see the stars making circles around Polaris as the Earth turns on its axis. Smartphone apps designed to capture star trails make this kind of photography very easy.

METEORS

Go outside during a meteor shower and take a sequence of photos. You can use an intervalometer or a timelapse function if the camera has one. This will automatically take photos for you. Sit back and enjoy the show. Check through your photos afterwards to see if any of them contain a meteor trail. This photo shows a bright Perseid meteor.

RESOURCES

Resources mentioned throughout the book are listed here.

NASA SPOT THE STATION
spotthestation.nasa.gov
Find out when to look out for the International Space Station in the sky.

STELLARIUM
stellarium.org
Free astronomy software for computers.

UPCOMING ECLIPSES (TIME AND DATE)
timeanddate.com/eclipse
Look up when and where eclipses will be visible around the world.

INTERNATIONAL METEOR ORGANISATION
imo.net/resources/calendar
Discover the dates and times when meteor showers will be active.

LIGHT POLLUTION MAP
lightpollutionmap.info
Check where has lower levels of light pollution to pick a great stargazing site.

GLOSSARY

10X50

Binocular sizes are given by two numbers and the letter x. For example, they may be labelled 6x30, 8x40 or 20x80. The first number describes how much magnification the binoculars provide. An 8x pair of binoculars will make everything appear eight times larger. The second number gives the width of the binocular lenses in millimetres. So an 8x40 binocular has 40 mm wide lenses. Wider lenses collect more light and produce brighter views.

APERTURE

The width of a lens or mirror in binoculars or a telescope. Aperture is usually measured in millimetres or inches. A 90 mm telescope has a main lens or main mirror that is 90 mm wide.

ASTROLOGICAL STAR SIGNS

Twelve equal sections of the sky that are each 30 degrees wide. The star signs lie along the apparent path of the Sun – the ecliptic. The zodiacal constellations inspired the names of the star signs. Zodiacal constellations vary in size, but star signs are all equal in size.

ASTROPHOTOGRAPHY

The art of photographing the night sky, including stars, planets, the Moon and Sun, deep sky objects, and special events such as meteor showers.

ATMOSPHERE

A layer of gas that surrounds the Earth, another planet or a star. The Earth's atmosphere contains air, which is the mixture of gases we can breathe. Other planets and stars also have atmospheres with different gases.

AVERTED VISION

Looking slightly to one side of a faint object so that it appears brighter. The centre of your vision has a small blind spot in dark conditions. Looking to one side of the object you want to see lets you use a more sensitive part of your eye.

BARGES

Storms on Jupiter that are long and dark in colour. Barges form in Jupiter's northern equatorial belt. Large barges can be seen with a telescope.

CONSTELLATIONS

Formal star patterns that are agreed upon by astronomers around the world. There are 88 modern constellations across the whole sky.

CONTRAST

The difference between darker and brighter regions of the sky. Light pollution makes the sky brighter, which lowers the contrast between the sky and the stars. Binoculars and telescopes can provide more contrast. Higher contrast is a benefit for viewing faint objects, such as nebulae and galaxies. Contrast may also describe how strongly colours appear on the surfaces of planets or between two members of a double star.

DEBRIS

The remains of something that has been broken or destroyed. The Solar System is full of bits of rock and ice that were left over from when the planets formed. These pieces of debris were left behind after objects crashed into each other, smashing them apart.

DISK

A flattened, round shape. In astronomy, the word 'disk' has several uses. Saturn's rings form a disk around the planet. The Milky Way (and every other spiral galaxy) has a disk of stars and gas where the spiral arms are found. When a planet is seen in a telescope, its face is called its disk.

DWARF PLANET

An object in the Solar System which orbits the Sun, is spherical or nearly spherical in shape, but shares its orbit with a large amount of debris. Pluto is a dwarf planet because its orbit is cluttered with rocky and icy dust. The Earth is not a dwarf planet because its orbit is mostly clear.

DWARF STAR

A type of star which is relatively small in size and has a relatively low brightness. The Sun is a dwarf star. Many other dwarf stars are much smaller than the Sun. By comparison, giant stars can be thousands of times larger than the Sun.

EDGE-ON (GALAXY)

Spiral galaxies are very wide and flat, with spiral arms sweeping out from the centre in a thin disk. When they are seen from the side, they are 'edge-on' to us and they appear as thin lines of light in the sky. We are seeing the edge of the disk, rather than the round face of the galaxy.

EYEPIECE

An accessory for a telescope which forms an image for you to see with your eye. Eyepieces magnify the view produced by the telescope. Eyepieces have a focal length, measured in millimetres. The smaller the focal length, the more magnification the eyepiece will provide. Eyepieces are placed into the telescope focuser.

FACE-ON (GALAXY)

A spiral galaxy seen from the top or bottom, so that its spiral arms are visible from Earth. Seen face-on, spiral galaxies appear round in telescopes.

FESTOONS

On Jupiter, festoons are strange, blue features that form near the equator. They seem to dangle from the northern equatorial belt. They are named after traditional chains of flowers or ribbons that are used for decorations.

GRAVITY

An invisible force which attracts objects together. Everything, including stars, planets, moons and even people, produces gravity. The Earth and the other planets remain in their orbits around the Sun because of gravity. Gravity also keeps us on the surface of the Earth. Gravity acts across very large distances. It causes stars and even galaxies to collect into clusters.

LIGHT YEAR

The distance light travels through space in one year. One light year is about 9.5 trillion (million million) kilometres, or 5.9 trillion miles! Light is the fastest thing in the Universe, but it still takes many years to travel between the stars. Space is so large that astronomers measure some distances in billions (thousands of millions) of light years.

LONG EXPOSURE

A photograph is taken when light shines onto a sensor. The sensor is exposed to light for a period of time. During the day, this is usually a small fraction of a second. In astrophotography, we need to expose the sensor for longer to collect more of the fainter light in the night sky. We use long exposures, such as one second, five seconds or even 30 seconds.

LUNAR

Things which are related to the Moon. For example, the Moon's surface is also called the lunar surface.

METEOROIDS

Any small object in space, that would become a meteor if it collided with the Earth's atmosphere, is called a meteoroid. Typically, this is a small piece of rock or ice. When a meteoroid enters the Earth's atmosphere, it becomes a meteor (also known as a shooting star). Some meteors travel all the way to the ground, and the piece of rock that reaches the ground is called a meteorite.

MOONGLOW

When the Moon shines brightly in the sky, its light illuminates Earth's atmosphere. While it is always much fainter than the Sun, the full moon can still make the sky much lighter, particularly in its surrounding region. This moonglow can hide faint stars and deep sky objects, making them difficult to see.

NEBULAE

A nebula (plural: nebulae) is a cloud of gas in space. Nebulae are many times larger than our whole Solar System. Stars are born inside nebulae, where the gas clumps together because of gravity. Stars also create nebulae when they die, either by exploding or gently expanding.

PLANE

A plane is an imaginary flat shape. Astronomers sometimes talk about the plane of the Solar System. This is an imaginary circle centred on the Sun's equator. The orbits of the planets are always close to this plane and so the Solar System looks almost flat from a distance.

SATELLITE GALAXIES

Just as planets have natural satellites called moons, many galaxies have satellite galaxies orbiting around them. The Milky Way has dozens of tiny galaxies orbiting around it. Other galaxies, such as the Andromeda Galaxy (M31), have bright satellite galaxies (M32 and M110) that we can see with telescopes.

SCATTERED

When light meets an atom, it is often scattered. Think of it as light bouncing off the atom. Different kinds of atoms scatter light differently. The atoms in Earth's atmosphere, for example, scatter blue light in every direction but they scatter red light in a direction similar to the one it is already travelling in. When white light from the Sun (which is made up of every colour) enters the atmosphere, the blue part is bounced all over the place, making the sky blue.

SHUTTER

A door in front of a camera sensor which opens and closes quickly to allow light to fall on the sensor during an exposure.

SILHOUETTE

The shadow of something which is lit from the other side. When a planet, or the Moon, crosses in front of the Sun, we see its silhouette. When a person stands in front of a bright lamp, their silhouette is seen. A silhouette is visible to you when you are in the subject's shadow.

SPIRAL ARMS

Spiral galaxies like the Milky Way are named for their appearance. Seen from the top or bottom, spiral galaxies have bright spiralling regions where stars are more concentrated. These are called spiral arms.

SYSTEMS (OF STARS)

Two or more stars form a star system when they are close enough that they orbit around each other. Two stars form a binary star system. More than two stars form a multiple star system.

TOTAL ECLIPSE

A total lunar eclipse occurs when the full moon is completely covered by the centre of the Earth's shadow. A total solar eclipse occurs when the new moon completely covers up the Sun.

INDEX

A

Achernar 35
Al Salib 60, 61
Albireo 25, 60-63, 130
Alcor 51, 130
Aldebaran 26, 78
Algieba 76, 77
Algol 68, 69
Alioth 51
Alkaid 51
Almach 68, 69, 131
Alnilam 79
Alnitak 79
Alpha Aurigid
  meteor shower 117
Alpha Capricornid
  meteor shower 60, 117
Alpha Centauri 35
Alphecca 64, 65
Altair 26, 62, 63
Andromeda . 60, 66, 68-70, 72, 73, 80, 131
Andromeda Galaxy (M31) 68, 69, 136
Antares 62, 63, 129
Antlia 78
Apennines Mountains . . . 90, 91
Aquarius 30, 31, 70, 71
Aquila 26, 60, 62, 63, 72
Arcturus 35, 52-55
Aries 30, 31, 68, 78, 80, 81
Aristarchus crater 90, 91
asterism 24-27
asteroid 112, 113
asteroid belt 112, 113

astrophotography 148-161
astrophotography
  accessories 152-153
astrophotography
  targets 156-161
Auriga 26, 56, 58, 68, 74, 76, 78-81
aurora borealis 142, 143
autumn equinox 32, 33
averted vision 41

B

barge (Jupiter) 107
Beehive Cluster (M44) 56, 57, 76, 77, 127
belt (Jupiter) 107
Belt of Venus 147
Betelgeuse 35, 78, 79
binoculars 42, 43
Blinking Planetary
  Nebula (C15) 72, 135
Blue Snowball Nebula (C22). 70, 71
Bode's Galaxy (M81) . . 50, 51, 137
Boötes . . . 50, 52–54, 58, 62, 64

C

C1 (Polarissima Cluster) . . 74, 75
C10 .68, 69
C14 (Double Cluster) .68, 69, 127
C15 (Blinking Planetary
  Nebula) 72, 135
C20 (North America Nebula) 133
C22 (Blue Snowball
  Nebula) 70, 71

C25 .... 56, 57
C41 (Hyades Cluster) . 80, 81, 126
C49 (Rosette Nebula) 133
C6 (Cat's Eye Nebula) ... 58, 59
Callisto 106
Camelopardalis 50, 56, 58,
.... 66-68, 74, 76, 78, 80
camera 151
Cancer 30, 31, 56, 57,
.... 76, 77, 78, 127
Canes Venatici 50, 52-54,
.... 56. 64, 76
Canis Major 26, 72, 78
Canis Minor 26, 56, 78
Canopus 35
Capella 26, 35, 78-81
Capricornus 30, 31, 70
Cassini Division 109
Cassiopeia .. 50, 58, 59, 66-70,
.... 72, 74, 75, 78, 80, 127
Castor 57, 78, 79, 131
Cat's Eye Nebula (C6) 58, 59
Cepheus 50, 58, 60, 66,
.... 68, 70, 72-74
Cetus 70, 78, 80
Cigar Galaxy (M82) . 50, 51, 137
Circlet asterism 70, 71
Clavius crater 90, 91
Coathanger 27, 73
Coma Berenices ... 52, 54, 55,
.... 64, 76
Coma Star Cluster 54, 55
comet 114-116
constellation 18-23, 30,
.... 31, 48-83
Copernicus crater 90, 91
Corona Borealis 52, 58, 60,
.... 62, 64, 65
Corvus 54
Crater 54
Crux 82, 83
Cygnus 25, 26., 60-63,

..... 66, 70, 72, 73, 125, 130, 133
Cygnus Star Clouds 73, 125

D

Deep Sky 120-139
Delphinus 60, 61, 72, 73
Delta Aquariid
   meteor shower 117
Deneb 25, 26, 60-63,
.... 72, 73, 125
Double Cluster (C14) . 68, 69, 127
Draco 50, 52, 58-60, 62,
.... 64-66, 72, 74, 75
Dubhe 28, 51
Dumbbell Nebula (M27) ....
.... 60, 61, 72, 73, 135

E

Eagle Nebula 63
eclipse (solar) 96-99
eclipse (lunar) 94, 95
ecliptic 31, 49
Enif 72, 73
Epsilon Lyrae 72, 131
equinox 32, 33
Equuleus 60, 70
Eridanus 78
Eta Aquariid meteor shower . 117
Europa 106
Evening Star (Venus) 103
eyes 40, 41

F

festoon (Jupiter) 107
filter 44, 89, 98, 101,
.... 103, 105, 107, 109
fireball 117
full moon 93

G

galaxy 55, 122, 123, 136–139
galaxy cluster 138, 139
Galilean Moons 106, 107
Ganymede 106
Gemini 26, 30, 31, 56, 57,
.... 76, 78, 79, 131
Geminid meteor shower . 76, 117
Giausar 74, 75
Globular Clusters 128, 129
Golden-Eye Cluster (M67) . 76, 77
Great Red Spot 107
Great Square of
   Pegasus asterism 68–71

H

Hamal 80, 81
Heart Nebula (IC 1805) ..68, 69
Helix Nebula (NGC 2392) . 70, 134
Hercules 27, 52, 58–60, 62,
.... 64, 65, 72, 128
Hercules Cluster (M13) ....
................64, 65, 128
Hyades Cluster (C41) . 80, 81, 126
Hydra 54, 55, 76

I

IC 1396 (Elephant's Trunk
   Nebula) 72
IC 1805 (Heart Nebula) ...68, 69
IC 1848 (Soul Nebula).....68, 69
ice giants 110, 111
International Space
   Station (ISS) 118
intervalometer 152
Io 106
ISS (International
   Space Station) 118

J

Jupiter 106, 107

K

Kemble's Cascade 66, 67
Keystone asterism 27, 64,
.... 65, 128
Kids (the) 80, 81
Kochab 58, 59

L

Lacerta ...60, 61, 66, 68, 70–72
Lagoon Nebula (M8) . 62, 63, 133
lenses 152
Leo 30, 31, 54–56, 64,
.... 76, 77, 138, 142
Leo Minor 50, 52, 54, 56,
................64, 76, 77
Leo Triplet 54, 55, 138, 139
Leonid meteor shower 117
Lepus 78
Libra 30, 31
light pollution 38, 39
long exposure 151
lunar eclipse 94, 95
lunar halo 144
Lynx . 50, 56–58, 66, 74, 76, 78,
Lyra ...26, 58, 60, 62, 63, 72, 131
Lyrid meteor shower 52, 117

M

M104 (Sombrero Galaxy) ......
.... 54, 55, 137–139
M11 (Wild Duck Cluster) ....
................62, 63, 127
M13 (Hercules Cluster) ....
................64, 65, 128
M15 (Pegasus Cluster) ....
.... 72, 73, 129

M22 . . . . 62, 63, 129
M23 63
M25 63
M27 (Dumbbell Nebula) . . . . . .
. . . . 60, 61, 72, 73, 135
M29 . . . . 60, 61
M3 52–55
M31 (Andromeda Galaxy) . . . . .
. . . . 68, 69, 136
M33 (Triangulum Galaxy) 68, 69
M34 (Spiral Cluster) . 66, 67, 127
M35 . . . . 56, 57, 78
M36 . . . . 78, 80, 81
M37 . . . . 78, 80, 81
M38 . . . . 78, 80, 81
M39 . . . . 60, 61, 66
M4 . . . . 62, 129
M42 (Orion Nebula) . 78, 79, 133
M44 (Beehive Cluster) . . . .
. . . . 56, 57, 76, 77, 127
M45 (Pleiades Cluster) . . . .
. . . . 78, 79, 126
M51 (Whirlpool Galaxy) . . . .
. . . . 50–53, 136
M57 (Ring Nebula) . . 62, 63, 134
M6 63
M65 . . . . 138, 139
M66 . . . . 138, 139
M67 (Golden-Eye Cluster) 76, 77
M7 63
M71 73
M72 71
M73 71
M8 (Lagoon Nebula) 62, 63, 133
M81 (Bode's Galaxy) . 50, 51, 137
M82 (Cigar Galaxy) . . . .
. . . . 50, 51, 137
M87 137–139
M92 . . . . 64, 65
M97 (Owl Nebula) . . 50, 51, 135
Mars 63, 104, 105
Megrez 51
Merak 28, 51
Mercury 100, 101
Mesarthim 80, 81
meteor 116, 117
meteor shower 116, 117
Milky Way 38, 49, 58–63,
. . . . 78, 79, 122–125
minor planet 112–115
Mintaka 79
Mirfak 66, 67
Mirfak Cluster 66–69
Mizar 28, 51, 130
Monoceros 64, 78, 133
Moon (the) 39
Moon (the) 39, 88–95
moon phases 39, 92, 93
moonlight 88
moons (Saturn) 108
Morning Star (Venus) 103
multiple star systems . . . 130, 131

N

nacreous clouds 145
nebula 122, 132–135
Neptune 110, 111
new moon 93
NGC 2392 . . . . 56, 57, 135
NGC 3628 . . . . 138, 141
NGC 5195 . . . . 52, 53
NGC 7293 (Helix Nebula). 70, 134
noctilucent clouds 144
North America Nebula (C20) 133
Northern Cross 25
Northern Taurid meteor shower .
. . . . 68, 117

O

open star clusters 126, 127
Ophiuchus 30, 31, 52, 62
Orion 19, 26, 78, 79, 116, 133
Orion Nebula (M42) . 78, 79, 133
Orion's Belt 79
Orionid meteor shower ....
.... 68, 116, 117
Owl Nebula (M97) ... 50, 51, 135

P

Palus Somni 90, 91
Pegasus ..60, 68, 70-73, 80, 129
Pegasus Cluster (M15) . 72, 73, 129
penumbra 94, 95
Perseid meteor shower ..60, 117
Perseus 19, 58, 59, 66-69,
.................74, 78, 80, 127
Phecda 51
pinhole projector 99
Pisces 30, 31, 70, 71, 80, 81
planetary nebulae 134, 135
Plato crater 90, 91
Pleiades Cluster (M45) ....
.... 78, 79, 126
Plough (the) 24, 28, 29, 130
Polaris 28, 29, 50, 58,
.... 59, 66, 74, 75
Polarissima Cluster (C1) ... 74, 75
Pollux 26, 57, 78, 79
Porrima 138, 139
Procyon 26, 35, 78
pupils 40

Q

Quadrantid meteor shower ....
....................74, 117

R

red light 41
Rigel 26, 35, 78, 79
Ring Nebula (M57) 134
Rosette Nebula (C49). 62, 63, 133
Rupes Recta 90, 91

S

Sagitta 27, 60, 72, 73
Sagittarius ....
.......25, 30, 31, 62, 63, 124, 133
satellites 118, 119
Saturn 108, 109
Scorpius 30, 31, 62, 63, 129
Scutum 62, 127
seasonal sky 17, 29
seasons 32, 33
Serpens Caput 52, 62, 64
Serpens Cauda 62
Seven Sisters 126
Sextans 54, 76
Sheratan 80, 81
Sickle asterism 54, 55
Sinus Iridum 90, 91
Sirius 26, 35, 78, 79
smartphone
 astrophotography 150
solar eclipse 96-99
Solar System 84-119
solstice 32, 33
Sombrero Galaxy (M104) .....
.... 54, 55, 137-139
Soul Nebula (IC 1848) 68, 69
southern stars 82, 83
Southern Taurid
 meteor shower 68, 117

Spica 54, 55, 138, 144
Spiral Cluster (M34) . 66, 67, 127
spring (vernal) equinox .. 32, 33
star brightness 34, 35
star charts 46-83
star cluster 122, 123
star colour 34, 35
summer solstice 32, 33
Summer Triangle 26, 62, 63, 125
Sun (the) 31-33, 96, 97
sun dogs 145
symbol (emission nebulae) .. 49
symbol (galaxy) 49
symbol (globular cluster) 49
symbol (meteor shower) 49
symbol (open star cluster) .. 49
symbol (planetary nebulae) . 49

T

Taurus .. 26, 30, 31, 68, 78-81, 126
Teapot asterism ....
.... 25, 62, 63, 124, 129
telescope 44, 45
terminator 88
Theophilus, Cyrillus
  and Catharina craters . 90, 91
Thuban 58, 59
Titan (Saturn) 108
Triangulum 68, 78, 80
Triangulum Galaxy (M33) . 68, 69
Trifid Nebula 62, 63
tripod 45, 153
true north 28, 75
Tycho crater 90, 91

U

umbra 94, 95
unusual sights 140-147
Uranus 110, 111
Ursa Major 19, 28, 50-54,
.... 56-59, 64, 66, 67, 74-78
Ursa Minor ...19, 28, 50, 52, 56,
.... 58, 59, 64, 66, 74, 75
Ursid meteor shower 74, 117

V

Vega 26, 35, 62, 63, 131
Veil Nebula 60, 61
Venus 102, 103
vernal (spring) equiniox .. 32, 33
Virgo 18, 30, 31, 54, 55, 64
Virgo Cluster 138, 139, 143
Vulpecula 27, 60, 72, 73

W

Whirlpool Galaxy (M51) ....
.... 50-53, 136
Wild Duck Cluster (M11) ....
.... 62, 63, 127
Winter Hexagon 26, 78, 79
winter solstice 32, 33

Z

zodiac 31
zodiacal light 146
zone (Jupiter) 107

CREDITS

All images © Shutterstock, and all diagrams created in-house except:

pp. 18-19, 30-32, 51, 53, 55, 57, 59, 61, 65, 67, 69, 71, 73, 75, 77, 79, 81, 84: Steve Evans

pp. 40-41, 43, 70 (Blue Snowball Nebula), 82, 88 (terminator), 89 (binocular views), 89 (telescope views), 91, 95, 97, 101, 103, 105, 107(t), 109, 115, 131(all), 133 (C49), 143, 157-161 (astrophography): © Tom Kerss

p.56 (M44): NASA/JPL/Stuart Heggie

p.66 (Kemble's Cascade): Wayne Young/CC 2.0

p.74 (C1): NOIRLab/NSF/AURA/CC 4.0

p.78 (M45): NASA/JPL/Space Science Institute

p.78 (M42): NASA/JPL-Caltech

p.99: Oona Stern/Science Photo Library

p.100: NASA

p.102: NASA/JPL-Caltech

p.106: NASA/JPL/USGS

p.108(t): NASA/JPL/Space Science Institute

p.108(b): NASA

p.113: NASA/JPL-Caltech/UCLA/MPS/DLR/IDA

p.126 (M45): NASA/JPL/Space Science Institute

p.127 (M44): NASA/JPL/Stuart Heggie

p.135 (C15): Judy Schmidt/CC 2.0

p.137 (M81/M82): Stephen Rahn / Public Domain

p.137 (M87): KPNO/NOIRLab/NSF/AURA/Chris Mihos (Case Western Reserve University)/ESO / CC 4.0

p.139(t): Nielander/CC 4.0

p.139(b): Fernando Pena/CC 2.0

p.145(t): Erik Axdahl Axda0002/CC ASA 2.5

p.146: ESO/Y. Beletsky/CC 4.0